D0146136

The Fading of the Greens

The Fading of the Greens

The Decline of Environmental Politics in the West

Anna Bramwell

Yale University Press
New Haven & London · 1994

Set in Meridien by Best-set Typesetter Ltd, Hong Kong
Printed and bound in Great Britain by Biddles Ltd, Guildford and Kings Lynn

Library of Congress Cataloging-in-Publication Data

Bramwell, Anna.
  The fading of the Greens : the decline of environmental politics
in the West / Anna Bramwell.
      p.   cm.
  Includes bibliographical references and index.
  ISBN 0–300–06040–8
  1. Green movement.   I. Title.
JA75.8.B73   1994
363.7'0525—dc20   94–21029
                                                                  CIP

A catalogue record for this book is available from the British Library

Contents

Acknowledgements and Explanation

THIS BOOK IS a sequel to my 1989 study, *Ecology in the 20th Century. A History*. In that work I argued that a key element in the development of Green ideas was the 'revolt of science against science' that occurred towards the end of the nineteenth century in Europe and North America.

In the 1970s and 1980s what had previously been a minority interest acquired a mass following and developed party organisations, which at one stage became major players in local and in national politics (especially in Germany). Yet the expected breakthrough into full participation in national politics did not occur, and at this level Green support has waned. Yet support for environmental causes has grown, and reached unprecedented levels; supporting such causes is politically respectable, yet at the same time has not been translated into 'normal' political behaviour. I have tried to explain this paradox.

Writing about this decline, and trying to explain 'why not' is depressing. My confident predictions to colleagues, editors and journalists that Greens would *not* do well in this or that forthcoming election where their prospective success had been hyped have not gone down well, even to people who would not dream of voting for any Green Party.*

---

* The minor troubles experienced by Britain's Green Party in 1993 – the leaving of its best-known leaders and the conversion of another to God – will not be further discussed in this book. They

Everyone likes environmental ideas, everyone sympath-
ises with them but practically nobody wants to vote for
Green parties. Why not? The earth is tied to a stake about
to be destroyed by wicked planet-killers and along the crest
gallops the cavalry . . . yet something is wrong. They are
mumbling to each other, their long hair blowing in the
wind, tripping over their sandals. A taunting and confident
group approaches from the other side of the hill, banners
red one side and black the other. They sail past the suffering
globe, intent on other matters – on Tao, on nuclear winter,
on global equality. The self-proclaimed saviours of the
planet seem to be interested in other things, in denying the
right to love, to differentiate,* to be human. Sometimes
they seem ineffective, sometimes plain mad. Sometimes,
but not always. Others galloping along the crest are concerned, truthful, serious, thoughtful. The trouble is, they
take so long to think about things, maybe because they
are in tune with the planetary rhythms, and they think so
self-referentially. And thus the earth remains tied to its
stake, with the cavalry beating each other up on the hill
crest.

People are reluctant to vote for Green parties. Protest
votes sometimes produce a blip on a chart, but the experience of Eastern Europe, where environmental issues virtually brought down the old regime but where no political
environmental movement developed, encapsulates in rapid
time frame two decades of party political ecology in the
West, with support for the issues not being matched by
support at the polls.

A factor in this was the success of the German Green

are a symptom not a cause of decline, and would be a mere blip
on the line of development if the underlying structure of the
party were strong.

* For example Peter Singer, in his *Animal Liberation*. Singer is very
worried that people might prefer lovable animals over unlovable
ones.

movement, which skewed the experience of other European Green parties. The decline of the German Greens helped drag others down. The German Greens declined because unification revealed flaws in their socio-political picture of the world; because the other parties took from them what could be absorbed by parties in the normal range of the political spectrum and left the unabsorbable; and because the German Greens were not dominantly Green at all. Ironically this may strengthen the party *qua* party, since the Greens may end up as the alternative permanent minority member of a coalition instead of the FDP (Free Democratic Party); but it will weaken the cause of the environment.

My earlier book was for me a voyage of discovery through unknown waters. Largely they were the waters of Europe, but, like Rimbaud's *Bateau ivre*, I sailed near the Indies and saw panthers. Upon publication, a controversy developed, focused on the immediate political situation of 1989 and the prospect of a Green surge in British politics. Among the political forebears of the Greens were members of the Nazi Party and right-wing sympathisers in Germany. I pursued this issue to see if there was a generically Fascist strain of Greenness, and concluded that there was not, that it was the German tradition rather than the Fascist tradition that was linked to the holistic and organic beliefs of 1930s ecologists of the right. Richard Gott of the *Guardian* responded with a swingeing attack on my arguments, which he described as perverse and dangerous, assuming that I wanted to encourage what he saw as a conjunction of nationalism, love of the land and desire for roots. Professor Andrew Dobson countered, in the *Times Higher Education Supplement*, that I had a habit of turning over stones and dismaying liberals with the revelation that all was not squeaky clean underneath.

In the fraught political situation of the time, several Green Party reviewers were upset by my ecological history, because by objectifying it I was removing it from their control.* However, once the political fuss had died down I had a series of thoughtful and complimentary reviews in France, Italy, Britain and the United States; and have been quoted as a source by a sufficiently wide group of authors to make me believe that the arguments have been accepted. The idea of a 'political box' caught on, and was the theme of a recent American work on American environmentalism. *Ecology in the 20th Century* is on most reading lists on the history of environmentalism, or its ethics. Once the dust had settled we were left with a core of hard research which suggested the following argument.

Political ecology as a distinct category arose in the late nineteenth century (which is why my book did not dwell long on the eighteenth century, the hunter-gatherer era, or other eras excluded by my thesis). I defined the roots of ecology as including organic geologists, as well as biologists and energy ecologists. This thesis was not an arbitrary one, but rested on the conundrum that political ecology incorporated scientific assumptions that became common around that time. (Of course this thesis was developed after studying and considering earlier eras, writers and scientific visions, but in a book of just 130,000 words and twenty-five pages of references, it was not possible to develop a negative critique of all other possible theories.) Ecologism crossed and recrossed political categories and boundaries over the next eight decades, with an underlying anti-capitalist critique that moved from left to right and left again between 1910 and 1960. It is now accepted that the alternative ecological idea can take many guises, that there

---

* I noted ruefully in Oxford that in their seminars historians of socialism at least did not have to fend off enraged foremen from Cowley motor works, getting up and saying 'Man and boy I been a socialist, and me father before me, and I never knew nothing about this Gramcsi bloke': discussing the roots of ecology to an audience of pained undergraduate green sympathisers made me realise the impenetrability of the partisan mind.

is an urban element in environmentalist passion; that it is the belief of a city dweller looking out and back rather than the belief of a peasant, and that the scientific roots of this political movement are crucial.

The post-war story was not told at any length in that book. It was touched on indeed only in so far as it was necessary to show the fully fledged phenomenon whose genesis I was tracing. In this book I have tried to bring the story up to the present day, and to explain the extraordinary variance between the ecological movement and its institutional formation.

I would like to take this opportunity to thank those who have commented on the text, including Andrew Dobson, Tim O'Riordan, Mark Almond and Ann Geneva. I would like to thank Carcanet Press, Ltd for permission to quote from C.S. Sisson's poem 'The Absence', and the *Times Higher Education Supplement* for permission to use material, originally published in different form, on Petra Kelly in November 1992 and on Arne Naess in July 1989. Many of the ideas for the section on the Italian Greens were taken from discussion with Mario Dani on his unpublished paper on the subject, and with Roger Griffin of Brooks University, Oxford; while exchanges of ideas with Ravi Rajan of Wolfson College, Oxford, who is working on forestry in India and the effect of Empire, provided fruitful stimulus. Any errors and awkwardnesses that may remain are entirely my responsibility.

Introduction

THE TITLE OF this book may appear to be deliberately provocative. It is not so intended. Talk of the decline of the Green movement does not imply that environmentalism is finished, nor are references to the religiosity of the Greens meant to imply that a mere wave of hysteria has passed over the world and is now ebbing away. Indeed, the apparent awakening of a care for the natural environment seems one of the most joyful events to happen towards the end of a dark century. By the decline of the Green movement I mean the end of the brief era of dedicated Green national politics. In this book I hope to chart the rise and fall of this phenomenon, not in terms of party history, elections and class mobilisation, but in terms of the internal coherence of their ends and their means.

To seek the ideological and philosophical roots of a movement is bound to be contentious among that movement's followers. The environmentalist constituency is now much wider than any party. When a movement becomes a political force, its slogans and claims familiar to everyone's ears, analysing its roots may seem merely an arcane academic exercise. But the past exerts its influence in ways of which we may be unaware. We may seek to relive history: we frequently seek to avoid doing so. The shadow of the past hangs over the Greens, especially in so far as there is a new Green history, one in which they perceive themselves as representing a victimised and secret tradition. And indeed there is an alternative Green tradition. Like many traditions of the twentieth century, it was developed in the nineteenth century. In *Ecology in the 20th Century* I argued that the ecological movement was founded in a scientific aberration of the late nineteenth century. I will be describing my theory at greater length later on; let me summarise it here. Firstly a scientific basis for ecological ideas was an essential precondition for the growth of environmentalism, which postulated a sickness, a wrongness about Western industrialised society. The diagnosis of this sickness included an awareness of finite resources, while part of the solution was to reorganise society so that resources could be used more efficiently. Ecologism, a term I coined in my last book to mean the doctrine of political ecology, remained a preserve of a small section of the European and American intelligentsia until after the First World War. It was taken up in the inter-war void by political activists who shared some of their preconceptions with right-wing movements: anti-capitalist, seeking a third way, obsessed with death and regeneration. This 'soft right' movement moved across the political spectrum after the Second World War, without really changing its preconceptions. It went 'soft left'. The story of the post-war weaving of ideological strands into the philosophy we know today as ecologism is a complex one, told later in this book. And because different nationalities interpreted ecologism differently, it is analysed according to country.

A further purpose of my brief examination of some of the bases for ecological ideas is to act as a framework for, and to some extent as a corrective to, the growing tendency for Green thinking to be presented purely by Green sympathisers.

A portrait painted by a loving member of a family will differ from a portrait painted by an outside artist. Those who write the history of an idea have to deal with the problem of cross-fertilisation, the problems of abstracting from the live and fallible and confused person to the morphological level, of distinguishing between the articulated and the disarticulated. Furthermore, ideas mutate. They shift from stratum to stratum, from decade to decade, responding to outside forces and inner impulses, changing with individuals as well as with 'world-historical forces'. Should one always believe the story told by the insider? People's perceptions of their motives and aims tell us something, but they do not tell us everything. Political writings have a dual purpose: they are descriptive and ascriptive. They present, and they persuade. Such accounts of an ideology written by the committed usually have an historical dimension. The ancestors claimed by today's Greens should not be accepted without careful examination.

When I began to think about ecological problems in the 1960s, it was a rarefied interest. My friends, many of whom were attracted by Maoist or other revolutionary ideas, or were active in student politics, seemed to me to miss the crucial danger point of that time's politics, which was the steamroller of Western- and American-dominated culture ironing out all values, whether rural or spiritual, on a worldwide basis. The ruination by roads of Britain's countryside, a countryside which I could still remember from its early 1950s innocence (no doubt a very relative innocence compared with that of 1900), the development caused by decentralist ideals, and intensive farming, seemed merely a symptom, but a tragic symptom, of that process. However, during the 1970s two events happened that made me rethink my ideas, and conclude that I really now had no clear-cut position on these issues and that I should not presume to judge between clashes of interest.

The first was my growing belief that any attempt to control resource use by means of state planning was bound to fail and to waste human effort and energy, which is also a resource, just as much as state-run projects misused physical resources. If attempts to plan for land and other resource use were bound to end in failure, why was this? It seemed that what was at fault was a policy which combined two contradictory ideas, to build, and not to build: to develop, and not to develop. Bodies such as the Mersey Docks Board were at odds with the national bodies who aimed to conserve. The contradiction was resolved by organising development under the aegis of local authorities, to a much greater extent than had happened in the 1930s. The result of this constraint was a purely arbitrary and artificial limitation on land resources, controls that could be and were ignored, corruption, and larger profits for developers. Furthermore, what went up was of poor quality, ugly and unloved, and not of a quality comparable to new European building.

My experience on going to live on a Herefordshire smallholding in 1972 was the second cautionary note. I mentioned this briefly in the introduction to *Ecology in the 20th Century*, too briefly perhaps, since my fumbling attempts to express my feelings were seen as condescending by one reviewer, and High Tory exclusivist by another. I wrote there of my love and appreciation of the outgoing yeoman farmer and his wife, but also of my sense of the grief and difficulty they endured, the hardships, especially for women, of the decentralised, low-resource-use way of life so attractive to urban ecological reformers. This family rose at five and broke the ice in their pitchers to wash, before starting up a cooking fire with twigs. In the 1950s, they still drove to Leominster market with their horse and cart, and lived off their own produce, buying in only tea and sugar. Their twenty-five acres were fringed with damson trees that turned gold early every autumn. A crab-apple orchard, now grubbed up to conform with European Community legislation, stretched up to a hilly field that had been a vineyard in the fourteenth century. Old oaks, cut down in 1982, marked the hedges, and there were twelve tall elms, which died in 1974. Outside the front door grew a twiggy pink rose, planted when the farmer came back from the First World War and rented this smallholding. The full-time work of a wife and two maiden daughters was required to support this unit, kill the mutton to melt the fat to make the candles, and scrape and clean out the holders day by day. . . .

Perhaps by perceiving the sadness as well as the values of such a life one is bound to attract the hostility of rural romantics, or of optimistic Kropotkinites. My struggle to live a self-sufficient life brought me into close contact with some of its hardships. Nonetheless, I retained a gut feeling about the value of the rural life and the countryside – shared, opinion polls tell us, by the majority in Britain and Germany, though not everywhere in Europe. And I live in the country still, because I am happier there. So this book, which looks at the phenomenon of the rise and fall of Green politics in Western Europe and North America since the 1970s, is not without sympathy for ecological values. It does, however, express a profound perturbation about the political means.

The Green movement in the 1980s has been very much a moving target. When I began a research fellowship at Trinity College, Oxford in January 1984, the Greens were still an esoteric interest, largely associated with the German Green Party. In an article published in late 1987, I could describe two existing attitudes towards the Greens: nice, or problematic but irrelevant. By the time my book on the history of ecological ideas from 1880 to 1945 was published (Bramwell, 1989) ecological ideas had gone in one bound from irrelevance to being accepted with an undue lack of criticism. When fears of global climatic change began to be taken seriously (a process that owed more to the accidental conjunction of Sir Crispin Tickell and Mrs Thatcher than to any new scientific evidence on the subject), national political Green movements were gradually eased into the background. So, unexpectedly and unfairly for them, the Greens achieved their highest level of public awareness and support on the crest of the 'greenhouse effect' wave, at the same moment that the strengthening of supra-national environmental agencies, the involvement of the EC in environmental directives and legislation, the international response to environmental degradation in Eastern Europe, all began to weaken the Greens' importance. The internationalisation of environmental issues will turn the national movements into a second-division affair. The future development of environmentalism will lie mainly in the development of accords, regimes and agreements between nation states.

However, the impetus of radical ecologism still perturbs the fabric of our time, and will do so for many more years, as its creed leaks into the vulnerable texture of mass consciousness, and as a generation of 1960s activists achieves its long march through the institutions. In 1992 Al Gore, the Vice-President of the United States, published a bestselling book on global Green philosophy that encapsulates all the rhetoric of the Green activist, and one of the practical results of his vice-presidency may be that the carbon taxes that were only a gleam in the eye ten years ago will soon be put into place in the United States.

What do Greens want? Since the days of the first *Blueprint for Survival* (Goldsmith, 1972), nearly twenty years ago now, ecologists have been thinking about new strategies for creating and implementing environmental programmes. There has been more thinking than concluding, and that is part of the pattern of the ecological mind, but still, schools of thought can be divided very roughly into reform ecologists and deep ecologists. (Division into Light and Dark Greens of the Green Party is a slightly different distinction.)

Reform ecologists believe that you can work through and with the existing system. They support recent EC directives on environmental hazards, and other environmental legislation, and feel that a supra-national body like the EC is a fairer and more effective means of enforcing such legislation than national governments. They believe in working with industry, and taking account of the need for economic growth. Essentially, they think that the current economic system has room for environmental concerns, and can factor them into their equations.

Deep ecologists are the more radical wing, and the source of the ecological passion. Like the early Christian churches, the exponents of the radical ecological creed have made their point and left their mark, but have failed to win widespread support because of their excessively puritanical approach.

Clothes-snatching by the established political organs of the Western world has meant that what can be reformed will be reformed: what cannot be, lying deeply embedded in our society, will not be. This is perhaps the worst possible outcome for deep ecologists, just as the improvement of working conditions in the twentieth century was the worst possible result for revolutionary socialists.

New social movements, postmodernism, New Age

Any book about contemporary politics needs to spell out its framework. Mine is descriptive, and not theoretical. Descriptive categories do not prevent authors from treating their descriptions as if they have a theoretical component, but in my view to offer a label is not to explain something. Explanation in social science is in any case a vexed area, and cannot, given the present state of the art, be seen as on a level with scientific explanation. Social scientists may hypothesise, suggest or perceive, but categories evolved by humans to describe human phenomena cannot be the subject of hypothesis, predictions, controlled experiments and conclusions. I apologise for labouring this well-worn theme, but some explanation is needed as to why I will not be approaching the Greens from the point of view of those who see them as a phenomenon on the lines of new social movements, etc.

The fact that the rise of environmentalism caught many sociologists by surprise is a blow to the predictive natural science model of sociology. What had seemed a middle-class movement, a minority sport, suddenly appeared to attract mass support, even if this support was not expressed in voting patterns. Various academic disciplines have been in the irritating position of having their predictions and explanations confounded, and of having to explain a phenomenon after the fact. Although many sociologists would reject the simple model of a social science able to explain society and understand causation on the lines of the natural sciences, their standing as commentators none-theless rests (outside their own discipline) on the assump-tion that their analyses can be trusted.

In Germany, for example, sociologists spanning a spec-trum from moderate left to pragmatic right were taken by surprise at the rapid change in the political status and acceptability of the Greens. Jürgen Habermas saw their attack on technology as potentially anti-democratic, while Niklas Luhmann developed an analysis of society which argued that communication between society and nature-outside-society is impossible. He feared that concern about environmental problems would turn into a moralising self-aggrandisement, and also create a climate of useless anxiety – useless because society was not able to act upon phenom-ena external to itself (Luhmann, 1989; Habermas, 1985). In their different ways, these two writers were resenting the appearance of an ideology, with strong moral claims, that did not fit in with their analysis of social forces. Habermas was assuming the presence of an anti-modern element in ecological thinking. Luhmann was concerned to protect his model of society as self-referential and unable to connect to objective externalities.

Other sociologists and political scientists have mulled over the place of environmentalism in a society affected by the conditions of post-modernity, environmentalism's potential role as a substitute for class-based politics, and indeed its extension of the critical function attributed to socialism (Harvey, 1989).

Modernism itself is a term used in various ways (Featherstone, 1988). Some writers conflate the modern with ideas stemming from the Enlightenment. These comprise atomistic individualism, progress, democracy, rationalism and the bundle of political instincts described in British political terminology as 'liberal'. Liberalism here means support for emancipation, freedom, equality and justice, and the belief that educating and training people to be citizens of the just state will help towards realising such a society. Critics of modernisation theory have pointed out that there is no obvious historical connection between the liberal Enlightened society and technology, urbanisation, industry or advanced capitalism. Mid-nineteenth-century Germany and late nineteenth-century Russia modernised 'from above', with a mixture of moderated autocracy above and an element of local democracy and autonomy below. Nonetheless, the belief in untrammelled human enquiry, the freedom to develop, to experiment, to change, that accompanies liberalism, has widely been associated with the development of the industrial civilisation of the West. Some writers have considered it a precondition of modernisation (Rostow, 1960; Gerschenkron, 1962).

A belief in rational scientific enquiry and in reason is also seen as essential to modernity. The epistemology of science requires a clear separation of object and subject, of observer and observed. The subject observes; the object knows its place under the hegemony of the perceiver. In turn, the observer would try to be objective, to depersonalise his or her role. Paradoxically, in order to remove the self from intrusion into the act of observing, the self became dominant, because it became isolated.

Some theories of modernity see this doctrine of the isolated self as a dominant theme in Western civilisation, but one which has at some time in the past eighty or ninety years been overturned. However, while there is a rough consensus on what modernism is, dating the arrival of postmodernity is more difficult. Again, developments in scientific thought can be used as a marker. Einstein's theory of relativity and the development of quantum physics threw doubt on the received idea that humanity could be the central observer. Particles behaved differently when they were observed. The act of observing affected the object. Theories of indeterminacy seemed to cast doubt on cause and effect, to strike at our intuitive concept of contingency.

In art, an attack on consensus, a fragmentation of presentation, seemed to occur from the turn of the century onwards. After the First World War, the process increased. Artistic and literary theory eventually caught up with the postmodernist experience. Old certainties vanished. Among these were the certainties we (confusingly) call modern in a different context: the belief by modernist architects like Le Corbusier that human behaviour could be controlled by suitable buildings, the hope that a Platonically just society could evolve through social scientists' prescriptions.

So one use of the term postmodernity is to imply the fragmentation and disappearance of the intellectual consensus of the last two centuries, the undermining of convincing themes of liberalism and socialism. Still, used in this sense, one may ask whether anybody but literary theorists and art critics need be terribly bothered. Who has been affected? Along with the Max Ernsts, the tradition of popular representational art has continued, as has vernacular architecture and genre literature. That there has been a loss of faith in the received wisdom of the unpopular modern among intellectuals may be true, but that there has been a widespread loss of faith in scientific truth and the possibility of justice among the broad masses is less evident. Crises in explanatory social theory may be said to be a very esoteric form of torture. Used in this sense, the concept of postmodernism is not a very relevant input to the growth of ecological movements.

Postmodernism's aesthetic aspect is perhaps its most important. According to the aesthetic theory, twentieth-century modernism was a response to the industrial and technological world, which was secular, nihilistic, technophile, scientific and rational. Postmodernism asserted humanity's need for religion, for spiritual and aesthetic values, for a creed. Faith in rationality was shattered by the 'new' science of quantum physics, chaos theory, indeterminacy. When abstract art denied the representational urges of the masses, pop music covers took its place. Similarly, as orthodox religion became secularised, occult and spiritualist movements gained support (Ashworth, 1980).

This perception of a void in cog-wheel culture matches one other theory of the causes of environmentalism. Stephen Cotgrove believes that material satiety and economic growth have produced a generation of sated youth seeking for values (Cotgrove and Duff, 1981; Cotgrove, 1982). This we might call the Holy Grail theory. Readers will remember that once King Arthur established his law-abiding kingdom, introducing and enforcing the ideal of chivalry, the knights became bored. Despite job satisfaction, enough to eat, and the cultural delights of Camelot to keep them happy, they still galloped away after the Holy Grail. One writer (Inglehart, 1981) has argued that when industrial society reached a certain level of development and prosperity, participants, especially the young, became dissatisfied with material values. Their dissatisfaction led them to search for ideas embodying spiritual values, not only ecological or environmental values but also the 'irrational' and occultist mystical ideas mentioned above. New spiritual movements are seen to have an anti-social and anti-political component that, according to the proponents of this theory, seem to merge surprisingly easily with other new movements that have been replacing orthodox radical creeds such as socialism. Part of the post-materialist package-deal includes 'soft' ideals such as pacifism and feminism.

Up to the early 1980s, surveys of groups to determine why they support environmentalism seemed to throw up significant results. When allowances were made for the temporary effect on respondents of issues taken up by the mass media, it appeared that although the relative importance given to environmental values in different countries differs, the level of environmental concern was surprisingly stable over time (Lowe and Ruedig, 1986a, p. 514). However, there are problems with relating 'post-materialism' to some level of satiety in industrial society. It is not easy to fit the timing of ecological movements in some countries into this pattern. Italy is one example, and the timing of environmentalism there will be discussed later. Generational theories also pose the problems of methodology outlined by Robert Wohl in *The Generation of 1919*; i.e., when does a generation begin and when does it end: can one really argue that all of those born after a certain date differ from those born before it? Still, surveys have suggested that a generational factor does seem to exist. In the USA, sociologists found a consistent link between youth, higher education and political liberalism, and concern for the quality of the environment (Van Liere and Dunlap, 1980, quoted in Lowe and Ruedig, 1986a, p. 514). Studies cited by Werner Huelsberg in *The German Greens* (1988) also show that age and a previously left-liberal political stance affect support for the Greens in Europe. In short, the younger, the Greener. The environmentalist movements in some East European countries seem to have followed on from processes or political liberalisation, although Czechoslovakia is an exception, since the Czech Spring of 1968 did not include an environmental protest, and environmental dissidents emerged after the revolution of 1989 despite previous bitter oppression.

To some extent, this apparent link between ecological movements, youth and liberalism emerges because enquiries into post-materialism were framed to include precisely the kind of values that appeal to a young constituency, with environmentalism being thrown in as one variable. Status, aesthetic satisfaction, esteem and love were among the other values examined. Later work on post-materialism focused on support for anti-nuclear movements, and in doing so ran into the problem of pre-defining the environmentalists it wanted to study. By doing so, by focusing on anti-nuclear movements for example, surveys failed to observe environmentalists who were inspired by other priorities. Recent developments in Germany, where the Bavarian CSU (Christian Social Union) has adopted conservative and rural Green ideas, show the danger in such pre-defining of your subject.

Another criticism of the idea of explaining political responses by generational theories is that the categories are too absolute, for when people's accustomed level of affluence seems to be endangered, different 'generations' will respond in the same manner: their concerns will shift back to preserving their standard of living (Lowe and Ruedig, 1986a, 'Review Article', p. 517). The argument that environmentalism is a by-product of economic affluence is problematic in that environmentalism demands a reduced standard of living, because it blames environmental destruction on the very process that (in a superficial sense, according to environmentalists) produces higher standards of living.

Other criticisms have come from environmentalists, who not unnaturally resent their beliefs and values being tidied away under some distancing classification or vague title. They believe that their views spring directly from the existence of environmental issues, from a perception of a growing crisis, from a fear of the inexorable and irreversible destruction of what they hold dear.

A cultural reaction against dominant trends in twentieth-century society does not have to relate to standards of living, but can surely be an objective criticism. Why should an industrial society suddenly wish to change its style? Societies are not passive subjects. They do not, for example, simply experience industrialisation from above (when industrialisation is forced on a society from above, it tends not to stick). Industrialisation grows out of societies that have the necessary preconditions, that are intrinsically rational, secular, mobile. Why should industrial societies suddenly stop wanting to be these things?

In their current state, sociological theories do not explain the timing of environmentalism, why it was the preserve of a few for so long and then became a mass movement. This kind of explanation is rather confined to a more old-fashioned political history. In the remainder of this study, three quite different ecological movements – in Germany, the United Kingdom and the United States of America – will be examined, using the experience of environmental problems in Eastern Europe as a comparison. Since environmentalism has become a global concern more quickly than anyone expected, I will look at the relation between global tactics and Green ends.

The choice of area is not arbitrary. There are dozens of Green parties today, but these three countries have historically been in the forefront of political and intellectual developments in this area – even though the first Green Party appeared in Tasmania. I looked at the reasons for this geographical skew in my earlier survey, and concluded that ecologism depended upon a body of scientific thought, disseminated throughout a literate and leisured class. Reformist and organicist ideas had also to be present, together with a dash of the Comtean heresy that the objective government of man by experts was possible. These conditions were fulfilled most closely in the three countries already mentioned. Writers working on the history of forestry and geography have found links between the three countries, while recent work on the urge to empire has discovered that for both Germany, envious of British expansion, and for Britain, the desire to observe diverse ecosystems and experiment with them was important (Grove, 1985). So one can add another component to the conditions necessary to produce environmental experimentation: the space to fulfil the experimental urge, the open frontier, the commune; and the experience of an alien and 'other' society which might seem to hold the secret of harmonious life. This ideal of otherness has been located for the USA, Germany and Britain in the East, especially in India, and, for the USA, in the American Indian. Surprising links between Germany and the United States have been found in the nineteenth-century conservation movement, while the existence of an organicist and vitalist tradition in German thought is by now a truism (Capra, 1988).

The historical survey contained in Part One is divided into two: from 1945 up to and including the 1960s, and the more recent era. This is because the 1960s come as a kind of natural break in all three countries. Before that decade, environmentalism was a fairly elitist, middle-class phenomenon. After that, it became a mass movement. This shift in constituency was accompanied by a shift in political identity.

This is a tender subject, and if environmentalism as a mass movement before 1970 was hampered by its associations with pre-war right-wing ideas and even Nazism, that is not for a moment to assert that post-war environmentalists were either of these things. One can, however, trace the immobilisation of alternative views after the Second World War to the victory, as it was presented at the time, of liberal, democratic and urban principles and of rationalism – the triumph of the modern world over barbarism. It was easy then to discredit ideas about eternal values, embodied in the countryside, especially, as elitist, individualistic, selfish, and so on. Of course, patriotism and 'roots' were called into play during the war, as much in Britain as in the Soviet Union. But as late as 1965, political commentators were linking devotees of 'muck and mystery' with sinister conspiracy theorists, anti-Semites and right-wing groups

(Thayer, 1965). Many ideas and values went under, including representational art. It was natural that the return of value-oriented art, humanist art, whether it be drawing classes at the Royal Academy or human-scale buildings, should have been reviled by the cultural establishment in a manner similar to the attack on Prince Charles for admiring traditional rural values ('Prince Charles love of thatched cottages like Hitler'), but it would be a pity if a fear of such attacks were to prevent one referring to this shadow from the past, which undoubtedly affected the timing of environmentalism after the war. A new generation had to appear, for whom talk of values and rural life was not tainted, before the movement could begin to grow. And even then it had to take the prevailing coloration of radicalism, and rely on two other factors powerful in the post-war political scene. One of these was the growing power of the international problem-finding organs, of global concern about health, population levels and economic development. The other was the growth of a scientific intelligentsia.

Science and ecologists

The loss of that faith in science which replaced faith in religion among the thinking classes *is* relevant to the growth of ecologism, which is itself a new quasi-religion that has values, a creed, a way of life and a priesthood. Nevertheless, this new movement depends heavily on science, although a reconstituted form of scientificity. Ecologism promotes holistic science and simultaneously attacks mechanistic or analytical forms of science. I will return to this apparent conundrum later in the book: the problem is whether the Greens should indeed be seen as a reaction to humanistic and Enlightenment values, as their past might suggest, because of their opposition to what they see as the dominant mechanistic and Cartesian paradigm of the last 300 years (Davy, 1984). Taking Green political claims at face value means accepting that Green politics are qualitatively different from non-Green forms, and involve a new political axis of nurturing versus exploitation, of reverence for life versus extermination. Green politics mean placing humanity on a level with other species and the biosphere, and even, if necessary to the survival of the biosphere and the current ecosystem, valuing humanity at a lower level. In this key issue the Greens seem to be displaying an anti-humanist element in their founding ideology that at first sight appears to be a long way from the values of the Enlightenment. Yet it is not so very far away. 'Nature' was seen as an important but benevolent force in Enlightenment thinking. It too must be included in the picture (Vereker, 1967). 'Natural rights' and the 'Rights of Nature' (Nash, 1989) are as much the child of the Enlightenment as is the concept of objectivity.

Ecologists today, in taking their values from nature, take up this belief. Natural right has become the unspoken justification for serious ecologism. And science enhances it. 'It is astonishing to note the deep trust they [the Greens] have in the natural science of ecology,' comments one writer (Langhuth, 1986, p. 64). It lends ecology 'almost religious characteristics because it provides . . . "knowledge . . . of all earthly processes"'.

Given this relationship between ecologists and natural science, the differences between Enlightenment and anti-Enlightenment ideology seem to blur. Greens do base their new religion on science. It may be bad science from time to time, biased, unreliable, alarmist, but nonetheless it is science. The vision of an interconnected nature grew from ecologists, not because no one had ever had such a vision before, but because they could develop the idea through observation and reason – the scientific method, in short. In turn, scientists from many other disciplines were to provide the background data for apocalyptic prophecies about population bombs, mass famines, nuclear winters, resource depletion, the greenhouse effect, and so on.

In one sense, therefore, Greens may be seen as centred well within the intellectual development of our time. The reformist wing certainly is; and so are left Greens, who have simply moved the rhetoric of alienation, exploitation and inequality from socialism to ecologism. In the section on Green strategy this process will be discussed further.

In this short introduction I have already talked about 'environmentalists', 'light and dark Greens', 'deep ecologists' and 'reform ecologists'. Before going on to consider the development of national Green movements, some discussion of this tangle of terminology may be useful.

In Europe the term 'Green' has swept into use to include new political parties, the broader social movements and single-issue groups. Die Grünen of Germany started the fashion for this term, providing this handy label almost by accident: the Rainbow List had room for an ecological party, and Green happened to be the colour left over, as in the case of P.G. Wodehouse's Brown Shorts. In the UK, 'Green' is the catch-all term, used to describe groups as diverse as Green Party activists and local nature conservationists. In Italy, 'Le Verde' is the name applied to the entire environmental movement, as well as to their Green List, while in Britain the Ecology Party changed its name to the Green Party in acknowledgement of this widely used term. There are Green parties in Czechoslovakia, but Green has not yet become the generic label. In Hungary, confusingly enough, Greens were called 'Blues', after the Blue Danube Circle of activists opposed to the building of the Nagymaros Dam. The Hungarian Green Party was virtually boycotted by the Blues in the free election of March 1990. The Bulgarian ecological movement is called 'Ecoglasnost'. In Romania, the many ecological groups and parties that have sprung up since the revolution of Christmas 1989 usually have the name 'Ecological' in their title, while the first and largest Polish environmental group was called the Polish Ecology Club.

Talk of 'Greens' in America will sometimes bring a puzzled look to the eyes of the listener. The word is slowly coming into use there, but is used rather differently from Europe. In the United States, 'deep ecologists' are contrasted to the Greens. Greens are seen as a party politically oriented movement, whereas the deep ecologists are fundamentalists, and 'environmentalism' includes all those who lobby or otherwise use political processes. Outside the USA, hardly anybody knows the term 'deep ecologist', but it represents something important enough to have a chapter to itself in this book.

Throughout most of the last two decades, most countries saw a distinction develop between radical and reform ecologists. However, in some countries the contrast between deep and shallow ecologists is virtually non-existent – in Italy, for example – while in others one wing of the movement predominates. In the Soviet Union deep ecology predominates, because the more democratic and less fundamentalist type of political dissenter tends to put other political ideals first.

Within the ecological movement there are vehement disagreements about names. Arne Naess, the founder of deep ecologism, devotes several chapters of one of his books (Naess, 1989) to defining ecology, ecologism, environmentalism and so on, but other members of the movement disagree with him. The terminology used by insiders already differs from that used by the general public, who do not always differentiate clearly between environmentalist, ecologist, deep ecologist and Green.

There are also subdivisions within each group in each country. Early in this century, environmentalism in the USA divided into two key camps, Preservationists and Conservationists. Preservationists wanted to manage the wilderness, in part for the benefit of humanity; to log the forests and protect the animal populations. Conservationists rejected this as being exploitative; they opposed dam-building, mass access to national parks, and interference with predators. The terms do not have the same resonance in Europe. In Germany, those who felt able to work within 'the system' were *Realos*, or realists, those who did not were *Fundis*, or fundamentalists. Deep and shallow ecologists, or dark Greens and light Greens, are the British equivalent.

Many of these bipolar categories refer to a division between radical and less radical: revolutionary and reformist. The distinctions are sometimes blurred towards the centre, but the two ends of the axis are clearly defined. The involvement of Greens with party politics has sharpened this distinction, and the effect of the prospect – however distant – of political power, has been even more divisive, with some sticking to principles, and some prepared to compromise. The growth of party Greenness, furthermore, has inevitably meant interplay with other parties – with Green Parties forced to consider broadening their platforms and worrying about coalitions; and has led to clothes-snatching, with the established parties stealing what is stealable and attractive from the Green parties. It has also meant various kinds of entryism. Entryism and furious reactions against it naturally affect the indigenous body of Greenness. But the radical–reformist distinction does not correspond to a left–right division. It corresponds rather to an equally significant political division: anarchists versus statists. Interestingly, *Realos* in the German Green Party include former members of the hard left, Maoists, and associates of the Red Army *Fraktion*, whereas *Fundis* are closer to the utopian socialists of the last century.

In the survey of national Green movements in Part One I will be using terms that conform to the usage of the country under discussion. Greens, the catch-all noun, will include those active in party political issues. 'Environmentalists' will generally refer to single-issue and non-ideological activists, who are concerned with problems such as air and water pollution or protecting wildlife. Ecologists will be deep, and interested in the philosophical basis of their ideas, so that shallow or reform ecologists will be the same as light Greens. As the scientific discipline of ecology is not involved in this work, no distinction needs to be made between political and biological ecology. It is, however, widely accepted that the premisses of the two are similar, and many writers, including myself, believe that the former could not have arisen without the latter.

PART ONE

A Historical Survey

CHAPTER ONE

The Nineteenth-Century Roots of Ecology

'ECOLOGISM IS A political box. It is a new box, into which many distinguished and important thinkers fit who fit only partially into other, better-known boxes' (Bramwell, 1989, p. 13). In 1989 this analysis of the roots of ecology was innovative in several respects. First, I argued that the nineteenth-century scientific revolution was necessary for the flowering of ecologism as we know it today, and that the naturalists and explorers of earlier ages were not ecologists. I concluded that an urban and scientifically literate society was necessary for ecologism to flourish.

By 1880, ideas about entropy (the dissipation of energy) had become known to a wide circle of literate and articulate people. Vitalism was partly a reaction to the entropy fear; and its main argument – that life shows an anti-entropic power of organisation – has not been satisfactorily refuted. Both entropic pessimism and vitalist optimism inspired ecologists: the first provoked plans to deal with forthcoming disaster, the second inspired the belief that nature had a force and purpose and value above that of random chaos.

The second root of ecology I identified was the development of a biology which was no longer centred on human beings, and which no longer distinguished qualitatively between humans and animals. This took place between the mid-nineteenth century and 1880.

The idea that ecology has historical roots will be strange to many people. Yet when you analyse what ecologists believe, you can find a clear link to mid-nineteenth-century developments, precursors of nearly all of their beliefs, and even similar solutions.

What do ecologists believe? They want to conserve finite resources. They see humanity and animals as equal, although not all ecologists are vegetarians or animal rights activists. Most of them prefer the country to the town, and rural life to life in urbanised areas, even though ecologists are not anti-technology as such. They believe that people should be rooted to one place, and so should things, since most trade is unnecessary and wastes resources. Self-sufficiency is good: decentralisation and bio-regions are the means to a more economical world. Exploitation of nature is bad, and the urge to exploit is often attributed to some bad cultural spirit – to that of Man as opposed to Woman, to that of the West as opposed to the East. Civilisation is seen in a negative way, as exterminatory, as destructive, as dominating. Ecologists think globally (though in the last century the prevailing Eurocentrism tended to affect even the most enlightened), and are pacifist, both ideals being essentially Western. They worry about the danger to primitive peoples from Western contacts, and frequently believe that untouched tribesmen are superior to Westerners in their way of life, and happier. They are concerned about environmental toxicities, especially those in the food chain, are for forest preservation and the planting of trees, retaining 'wild' genes as against monoculture and excessive land clearance. Some ecologists are concerned with some issues rather than others, and different national cultures have had different emphases.

There are, of course, supporters of the Greens who are totally uninterested in matriarchy or vitalism. To many people, the situation seems simple. Nature and animals should be preserved, pollution should be cleaned up. Indeed, of the multiple package I have just outlined, two ideas today dominate: that air, water and soil should be as unpolluted as possible, and that resources should be used carefully and recycled.

However, human nature being what it is, these issues have not been adopted in isolation from the justifying ideology. And indeed, the choice of a clean environment and minimal resource use is a complex one, demanding much calculation and effort. As just one example, take the recycling of glass bottles. This is one of the easiest forms of recycling, and has attracted considerable public support. Yet it can hardly be argued that the major component of glass – sand – is scarce, and the energy used to melt down old bottles, clean and reuse them is also considerable. The choice to recycle is being made on an emotional level, and it is at that level that complex ideological principles have sunk in. To quote Anne Chisholm, in her 1972 book on ecologists, 'Here was a new morality, and a strategy for human survival rolled into one.' The cutting edge of environmental policy was until recently being developed by ideologues. Without that fact would people have the confidence to support environmental issues, with their aesthetic and indulgent aspect? Most Westerners today doubt the old certainty that the interests of the natural world must be subordinated to those of humanity. Even though we do not live according to ecological ideals, they have taken on the force of a moral standard. A wry apology is the usual response when we breach these ideals, seldom an impassioned defence of our actions in so doing. To have reversed the idea of the acceptable and provided justification for so doing can only be the work of a vigorous ideology.

The 'package-deal' of ecological ideas that appeared in Europe in the nineteenth century mimics that of today in a surprisingly complete way – even to the belief that matriarchal societies are superior to patriarchal ones, more peaceful and nurturing. The bundle of concepts shares an anti-capitalist heritage with the ideas of today's ecologists, and was similarly dominated by German-speaking thinkers (Johann Bachofen, nineteenth-century Swiss author of works on Bronze Age matriarchy; Rudolph Steiner,

Austrian inventor of biodynamic agriculture; Ernst Haeckel, the German anti-anthropocentric biologist).

The ecological world-view evolved in the nineteenth century out of a scientific development which meant that the idea of finite resources and energy scarcity became familiar to a wide body of educated people. The Second Law of Thermodynamics showed that energy was always dissipating. This was an exciting idea, but a depressing one which furthermore seemed to be contradicted by everyday experience in one crucial area – life itself. Evolution had demonstrated that living beings were increasingly complex and successful. Amoeba to man – child to man – was there some fundamental spirit at work here that opposed entropy, as the process of dissipation was called? The 'life force' and pantheism, helped to fill this explanatory void, and reinforced the nature-loving values of the thinkers concerned.

Disillusionment with Western progress appeared then as now. Western ideals were seen by this small minority of scientists as a destructive force, destined to triumph over the small nations and cultures of the world. The writings of organicist geographers of the period, for example, show a protective approach to marginal tribes which is partly paternal but partly the result of seeing earth's populations as part of a massive experiment, one we had no right to interfere with. There was a story being told: once upon a time people lived in caves, in huts, in tents. . . . We could learn from these living examples of parallel universes, and they represented alternative ways of living that were free of our errors and our guilts. To pollute them with Western ideas represented several kinds of wrongdoing: not only damaging the future of the tribes themselves, but also the psychic needs of the Western observers.

One of the themes of this study is the link between science and political ecology, and in particular the role played by science in lending conviction to what is at root a value-saturated creed. For ecology, science appears able to answer the question awkward for all creeds: 'where do the values come from?' It can claim to be an objective description of human needs and welfare, to have objective forecasting capacities. Given that humanity lives on the planet as it is, to forecast the planet's imminent disturbance can be taken *prima facie* as something of which we should take note. It is an advanced Green who will say 'so what', and a daring citizen who will doubt the scientific forecast.

It has become a truism to say that science became the 'new religion' in the nineteenth century, but that does not mean it is an inaccurate statement. The scientific vision of this period forms the starting point for political ecology in the twentieth century. The ecological movement is distinctive in that the countries where it is strongest all think that they invented it. This gives the movement a national dimension, which at first sight seems to parallel that of those political parties which have developed nationally, and seems to distinguish it from religious movements, which cut across national boundaries. Nonetheless, though retrospectively subject to nationalist interpretation, political ecology developed among a supra-national group of scientists, connected by journals, correspondence and peer pressure. Ecology resembled a religion in that scientists became 'converted' as a result of a new vision of the world and its life cycle. A consistent and thought-out body of ideas was pursued with passionate emotion. This process should be distinguished from the fierce disputes which commonly occur between rival scientists; the implications and arguments of ecology lay *away* from the mainstream interests of most of those involved.

Early ecologists were politically marginalised. They stood out against the power and manipulative ability of the time, the capacity to control, coerce and discipline hordes of labourers, the 'telegrams and anger' that E.M. Forster's heroine attacks in *Howards End*, the hierarchy and the inequality, the centralised wealth and dominance of the

West. These are alien to the scientific project; and the belief that characterises the left-ecologist today – in a more egalitarian world, with national and imperial boundaries overthrown and full participatory democracy replacing elitism – reflects the global network of the scientific guild. The biological and the earth sciences as we know them today developed from the momentum of the late nineteenth century. In Germany, France, Russia, Britain and the USA, plant science and zoology looked at organisms in their environment, and studied the feedback between energy input, nourishment, who ate, and who was eaten. As early as 1900, ecology was described as having important implications for mankind, and although plant scientists had vehement disagreements about the theory behind ecology, this vision of man as part of the biological life cycle was a constant. Geographers also began to see the earth as an 'organism', almost as a being in itself. The earth as organism was a metaphor, but an important one, signifying a shift to a more holistic vision.

These sciences were dominated by Germany, Britain and the USA. 'Organic' geography came from scientists such as Ernst Fischer, who predicted that untrammelled economic growth would use up natural resources and pollute the earth irrevocably, and from British geographers like A.J. Herbertson, who saw the earth as a living organism. The new science of forestry was accompanied by a mystique of the forests, which attributed almost everything bad in the past to deforestation. Interestingly enough, when the source book on forestry science was translated from German to English for an American publisher, the more mystical passages were omitted, while Gifford Pinchot, founder of American forestry science, studied the subject in Germany, but became associated with a resource management philosophy that was 'progressive' rather than mystical.

Ernst Haeckel, the biologist, coined the term *Oekologie*. Ethology, bird studies and plant science were fields led by Britain. Setting up wilderness areas, under the name of national parks or nature reserves, started in the USA in the mid-nineteenth century, while East Coast naturalists were joined by Californian ones by the end of the nineteenth century. The 'climax' theory of biology, which believed that nature's final aim was a stable and diverse ecosystem, was coined by an American biologist, Frederick Clements, and pursued by British plant scientists later on (Worster, 1977).

Patrick Geddes, Scottish architect and visionary, now hailed as an early 'Green', probably coined the term 'human ecology', while Lewis Mumford, the American architectural critic and town planner, prophesied that bio-regions would replace existing political units. Professor of chemistry Frederick Soddy in Britain and the Austrian Popper-Lynkeus called for a more egalitarian society, to distribute and ration resources more efficiently.

French and Russian scientists were also important in the development of scientific ecology – indeed, the term 'biosphere' was coined by a Russian *émigré* Geoffroy St Hilaire in Paris in the 1920s. St Hilaire was a zoologist who got near to *Oekologie*, and Kropotkin, the anarchist, calculated figures for self-sufficient and decentralised farming. Danilevsky, a biologist, was influenced by Darwin and obsessed by an ecological vision of the life cycle. In France and Russia the numbers of scientists involved seems to have failed to reach critical mass, but in Germany, the USA and Britain a stronger tradition of ecological ideas developed.

What of the non-scientific element in ecological ideas, the high value attributed to animals and nature, the disaffection with towns and trade? David Pepper (1985) has argued that this set of values emerges from the Romantic movement. Certainly, an element of nationalism, which was an important component of Romanticism, can be found among many chroniclers of the global movement, who are prone to concentrate on finding the roots of ecologism in their own national traditions. German ecologists point to the holism of Goethe, Paracelsus, Nietzsche and others (Bahro, 1986). Americans see Emerson, Thoreau and Whitman as the first, or at least the most important, celebrators of man and nature (Worster, 1977), while in Britain Matthew Arnold and the Romantic movement have been suggested as the origins of pantheistic nature-feeling (Keith, 1975; Pepper, 1985).

Certainly, the countryside plays a special role in the psyche of the three nations, although for different reasons. The North American wilderness signifies a god-haunted and undespoiled paradise to a people whose monotheistic religion sadly lacks a suitable desert. Pantheism in Germany reached its best-known expression with the *Wandervoegel*, the young students who took to the woods and fields before the First World War. The role of rural life in Britain has been widely studied, and perhaps needs no further stressing here (Gould, 1988; Marsh, 1982).

Finally, ecologism is a product of urban life. We do not call peasants ecologists, nor Indian tribesmen (though there is a school of thought that holds the hunter-gatherer to be the *real* ecologist (Oelschlaeger, 1991), despite the fact that hunter-gathering led to deforestation and other vast changes in the ecology, and the damage wrought by pastoral tribes today). The lovers of the wilderness, hiking and back-packing through America's national parks, go back to all the appurtenances of advanced technology at the end of their vacation. The educated classes who are the backbone of the environmentalist movement everywhere are a luxury of a wealthy civilisation that can afford to have up to one-third of its citizens kept out of employment until they are in their early twenties. And among urban nations it is the 'Northern White Empire', as Johann Galtung calls it, the 'Protestant triangle', as another writer describes the area covering Northern Europe and San Francisco, that is still dominant (Galtung, 1984; Martinez-Alier, 1987). The tendency to interpret impending ecological disasters as a punishment for human selfishness (Lutzenberger, 1986), the claim that preserving finite resources is a moral duty, the ecologists' attack on laying up earthly riches, may indeed be a residue of the Puritan ethic that once dominated much of this heartland. And whatever the degree of internationalisation of environmental issues, it is these countries that have been foremost in seeing problems, proposing solutions, reaching accords and, most recently, suggesting that the developing countries swap their debts in exchange for forest preservation and other environmental improvements.

This was the proto-ecological time-frame, and the organicist scientists of 1870–1920 were the seeds. The period up to 1970 was the seedbed, and we shall now look at key developments in the USA, Germany and Britain during this time.

CHAPTER TWO

The Northern White Empire

The USA 1945 to 1970

THE USA IS an important contrast to the European experience of environmentalism. North America has wildernesses. It has ancient forests uncut by human hand since the dawn of time. Until the 1930s, it had great savannahs which had never seen the plough. Even today, with the wildernesses tamed, ranged and trailed, it has the luxury of vast areas of land which have never been settled or farmed, and its land use in settled areas is profligate. Most of Europe, in contrast, has been settled and farmed for thousands of years. The chestnuts and oaks of the north Mediterranean littoral were cut down by Homer's time. The arid landscapes of today's North Africa once supported fertile fields of grain. Even though England is perhaps unique in its farmed landscape, the preservation of what has been lovingly created over hundreds of years by human hands poses a different set of conceptual and ethical problems from the preservation of wilderness areas.

The North American approach to its land has thus differed from that of Europe. A love of wilderness existed alongside an exploitative attitude to land and natural resources. The settler drive in the West met the shortage of water. In pursuit of the American dream of homesteads, of the Jeffersonian small farmer, federal laws encouraged small-scale arable farming in the arid lands of the Great Plains and the still more arid lands of the West. The result

– exacerbated by drought, but nonetheless caused by over-
ploughing and over-grazing – was the Dust Bowl of the
1930s, the failure of intensive pastoral farming, and the
chronic water shortages of the American West today. One
legacy of this experience which has pervaded the thinking
of American environmentalists is a strong distrust of private
ownership.

In 1945 nobody would have prophesied that environmentalism would become a major political force in North
America within two decades. Nor that environmentalism
would join the other exported political phenomena that
influenced Europe, nor yet that, in a pro-business culture,
strict liability environmental legislation would turn around
corporate thinking, and, through multinational behaviour,
enforce stricter environmental rules on European business.

And yet concern over land policies of various sorts had
dominated political thinking since the late nineteenth century, culminating in the intense wave of pro-government
ownership under Roosevelt's New Deal. The two main
forces had been the Preservationists and the Conservationists. Preservationists believed that the federal government
should own and protect wilderness areas, but allow controlled grazing, logging and access. Preservationist nationalism wanted the wilderness areas to serve the country.
Conservationists wanted the wilderness to remain untouched: it had value as and of itself. The Conservationist
cause seemed to go under with the building of the Hetch
Hetchy Dam, in Yosemite National Park, which broke the
heart of the most pantheistic of all nature-lovers, John
Muir.

Both wings of the wilderness movement, however, assumed that state ownership was the only way to preserve
natural landscapes. The US government had already played
the major role in allocating land to settlers. To reserve some
land and protect it from settlement was a natural consequence of its powers. It may be that in the long run the very
frontier mentality that inspired the drive to the wilderness

and an appreciation of it was an obstacle to its survival. It prevented the development of a *private* conservationist approach and encouraged the ideological conflict between state-preserved wilderness and privately exploited soil. A more conservative and long-term mentality was needed for private ownership to preserve a landscape.

Exploitation was attributed to private ownership in general, and opposition to private ownership of unspoiled territory was a natural reaction to the spectre of large corporations rampaging over the face of the country. So powerful was the opposition to private ownership of wilderness areas that today over 40 per cent of American land is owned by federal and state agencies, which own 33 per cent and 9 per cent respectively. One interesting example of the power of this historical myth is the fact that American environmentalists today call for the nationalisation of electricity plants (which are privately owned in America), despite the evidence from state-owned plants in Western and Eastern Europe that pollution is worse in state-owned utilities. This bias against private ownership and market forces is prominent in environmentalism generally, in both its right and left wings, and was especially strong in the United States scientific community, with its belief in planning and its concentration on native experience. The Californian environmentalists working out of Berkeley at the turn of the century possessed a mutated version of the frontier mentality. This was their playground, their experimental station, and it was their duty to prevent it from being ravaged by human settlement and development (Smith, 1988). Yet the direction in which Preservationism had developed had already demonstrated the dilemma: the overriding aim of the state tends to the protection and strengthening of the nation. The ideology of Conservation conflicts with the functional duty of the state, to utilise and exploit the nation's resources, and to protect them.

However, perhaps due to the growing corruption and size of the corporate power structure of inter-war America, the radical position was increasingly to support an extension of state power against the tradition of private property. Aldo Leopold's *Sand County Almanac*, published shortly after the Second World War, established the author as the voice of the environmentalist conscience, a pantheist successor to John Muir. Leopold had trained as a forester in the Pinchot school, and his thinking had been formed by Pinchot's vision of the trained cadre of super-foresters. He was already an authority on the management of wilderness areas, and his 1933 book, *Game Management*, was described by Donald Worster as 'the bible of the wildlife profession' (Worster, 1977, p. 271). It expressed his Preservationist philosophy that, 'Effective conservation requires . . . a deliberate and purposeful manipulation of the environment' (quoted ibid., p. 272). This involved careful management of animal life, killing predators to preserve herds of deer, and working for an increase of those products of the wilderness that could be used or consumed by man.

In 1935 Leopold visited Germany, and was converted from his Pinchonite approach to forest management to a more holistic one (Trommer, 1990). In 1949, when the *Sand County Almanac* was published, Leopold seemed to have changed his mind, and swung round to the Conservationist camp. He argued that even predators deserved respect: wilderness was no longer a human-centred ideal. He now offered an appealing rationale for maintaining areas of wilderness, stating that the American wilderness was 'the source of a unique American way of life' (Baden, 1989). The free enquiring spirit, the independent individualism and intolerance of idle folk that characterised American democracy resulted from the wilderness experience. It followed that the wilderness should be preserved to protect this cultural spirit.

This is in intriguing opposition to the popular ecologists' belief that the untrammelled exploitation of nature is a result of the independent, fearless enquiring mind, and the awareness of an apparently unlimited wilderness outside the back door that characterised the frontier mentality in the United States. This did not prevent it being an inspiration to the next generation of environmental activists. Leopold offered a new formula: the idea of the rights of nature, the right of the ecosystem to be preserved. He argued that precisely because man was part of nature, he must develop a new ethic, an 'ecological conscience' (ibid., p. 42), a phrase that was to become famous. With this, Leopold introduced the concept of man–animal equality. Where there was a conflict of interest between humanity and the natural world, it should not be assumed that nature should give way. Right and wrong were concepts that should be applied to nature as well as to society; rightness meant preserving a stable 'biotic community'; wrongness was disturbing the integrity of that community.

This argument dodged some problems that were to recur later in the 1970s, when activists began to try to incorporate the rights of nature into serious legal thinking. If, for example, man was part of nature, how could he have an ethic *towards* nature? Where did the ethic come from? If it was already embodied in nature, then why did we, already part of nature, have to develop it? Was not the idea of rights an analytical fiction, incompatible with the wholeness and all-inclusiveness, the benevolent cruelty, of nature? At what stage did a biotic community earn the right to have its stability preserved? Could there be rights without duties, without consciousness?

Despite these definitional problems, as a statement of intent Leopold's concept fired the imagination of a generation. His analysis of the place of wilderness in the national psyche had something for everyone: for the independent mountainman and the back-packer (though not the hunter, since he opposed killing predators). Perhaps few of those who read his *Almanac* really believed that wilderness would preserve the spirit of American democracy. One fan, though, was Richard Nixon who, according to John Baden's survey of American environmentalism, cited Leopold in a speech to Congress in 1971 (ibid., p. 49).

Leopold's writings had a folksy quality that appealed widely. Here was the Gospel of Nature, a sermon in stones and sandflies, home-grown wisdom, a call for a better quality of life at a time when small-town America was seen as under threat.

Leopold's natural wisdom was optimistic. In the 1960s a new element that was at once defensive and apocalyptic entered environmentalist thinking with the publication of Rachel Carson's *Silent Spring* (1962), which claimed that pesticides and other forms of pollution were getting into the food chain of birds, fishes and animals, and killing off many species. *Silent Spring* was erudite, readable, and was written by a woman whose wealth of experience as a natural scientist made her picture of a devastated natural world convincing, and who wrote with passion and fear. Its message was persuasive, focusing with that worrying clarity achieved by visionaries on a single issue, tracing its effects throughout the ecological system, and providing a guilty party – the pesticides industry. The book was a bestseller, and enormously influential.

Contemporaneously with this fearful picture arose the idea of the population bomb, a metaphor feeding upon the anti-nuclear movement of the 1960s. Paul Ehrlich's famous book of this name (Ehrlich, 1968) was co-published by the Sierra Club, an established and respectable conservation group. While solutions to the predicted population explosion differed – one group wanted to sweeten the pill for Third World countries by an egalitarian approach to resource use, while another supported compulsory sterilisation – the vision behind the campaign emphasised the destructive nature of human beings.

This new approach, which was prepared to abandon humanistic traditions, and put nature first, did not succeed without opposition. The Preservationists had opted for man against the wilderness when it came to the crunch (the Hetch Hetchy reservoir was finally built to meet the demand for water after the San Francisco earthquake of

1906); indeed, they had not been aware of a conflict. Marxists, who were increasingly interested in environmentalism and saw it as an extension of their humanist concerns, also rejected the idea of putting human interests behind those of nature.

Although for many observers environmentalism hit the public eye towards the end of the 1970s, it was the 1960s that was the decade of serious environmental legislation (McCormick, 1989). While there was a series of pollution disasters during this decade, there were no more than in the ten years previously. However, the American public was now sensitised to disaster, and attuned to lobbying persistently for its ends. The insecticide DDT was banned in 1965, and informal agreements were reached between the Public Works Committee and the most polluting industries to try to reduce toxic emissions and waste. This was clearly inadequate to halt or clean up the growing pollution, and lobbying began with a view to changing the law. John Baden (1989) has calculated that campaigns carried on in the 1960s resulted in about thirty-three environmental laws being passed between 1969 and 1977.

Environmentalists used common law concepts, whereby the citizen could sue pollutors for injuries received, and a body of case law built up. The problem was how to establish damage – in Britain a plaintiff in a civil action has to be involved, to have 'standing'. An important case of 1965 held that a citizen could be harmed or potentially harmed by a threat to scenic values, and the Sierra Club was able to bring cases under this ruling on behalf of its members (Baden, 1989, p. 53). However, a 1969 conference of environmental lobbyists decided on a statutory approach, and President Nixon signed the National Environmental Policy Act on 1 January 1970, the first measure to enforce environmental impact statements as a precondition of major federal projects.

The history of the Sierra Club can stand in miniature for the change in environmentalism from minority interest to mass movement. Founded in the late 1890s, the Sierra Club conformed for most of its life to the conservative elite model of conservationist groups (as with Italia Nostra, the Italian group founded in the early 1950s, and the British National Trust). Originally designed to protect the Sierras, it concentrated on arranging educational tours and visits to the national parks there, and lobbying at state level. Membership was low. In 1952 David Brower became director, and set about making the Sierra Club into a mass membership organisation of national scope. By 1956 membership was 10,000, after a successful campaign against the building of a dam in Echo Canyon, Colorado. By the time Brower left in 1970 there were over 100,000 members. Brower then founded the American Friends of the Earth. As Friends of the Earth developed into a powerful lobby group, Brower moved to support Earth First!, an eco-sabotage movement. The growing acceptance of the need to legislate, the growing body of proof that pesticide and other chemical companies needed controlling, simply added fuel to the fire of environmental activism. Not only the anti-nuclear movement of the 1960s but the anti-Vietnam war movement radicalised environmentalism in America.

One very successful film of the period was *Easy Rider* (1969), in which two attractive young hippies take to the roads on their powerful motorcycles. They experience the beauty of the untouched wilderness, bathe naked in mountain lakes with free-spirited young women, and live by drug-smuggling. Eventually one is killed by a red-neck Georgia farmer – a *farmer*, resenting the nomad incursion into his territory. Something significant about the 1960s experience lies behind this, once the Rousseauist trappings have been stripped off. It embodies several of the themes that were to dominate fully fledged American environmentalism during the 1970s and 1980s. There is the dream of the innocent wilderness with free and healthy sex thrown in; the self-pitying identification with apocalypse; the anarchic undertones; and finally the sense of death and regeneration that has characterised the radical movements of our century.

Germany 1945 to 1970

Today, the German Greens are the best-known Green group in Europe. Yet environmentalism in Germany was virtually a non-issue until the late 1960s. It then had a strange and chequered history, which bears very little resemblance to the history of Green parties elsewhere. Because of its size and (until recently) growing political power, there has been a natural tendency to take the German Green Party as a model, as the type to which other Green movements would conform. But ecologism in Germany differs from most European Green movements in its complete break with pre-war ecologism, and then in a post-war development that placed ideology above issues. That is perhaps why when environmentalism did appear it was characterised by two strikingly different political forms. The first was conservative, a reaction to the *Wirtschaftswunder* era by a Christian Democrat politician, Herbert Gruhl; the second was the growth of the so-called *Bürgerinitiative*, or citizens' movement, which used local pressure and referenda as a means of preventing the building of nuclear power stations, and which focused on pacifism and anti-nuclear power. Even this left environmentalism of the late 1960s and early 1970s was largely derived from the American New Left, though the fact that the New Left was itself inspired by former German philosophers like Marcuse presumably made the adaptation easier.

To some extent, then, a review of this early period in Germany is an attempt to answer the question, why was the development of ecologism delayed and, when it did appear, fractured? One reason is precisely the presence of a strong ecological tradition under the Nazis (Bramwell, 1985). After the war, any talk of holism, or a love of nature that adduced certain values from nature and strove to adapt humanity to those values, was suspect. It was also perceived as irrelevant. The reconstruction of Germany from 1945 on was remarkably fast and was carried out without backward looks. Resentment of industrialisation seemed meaningless to the great majority of inhabitants in a country in ruins; so did hostility to materialism when civilians in their millions were experiencing dislocation, hunger and disease.

The tainted philosophical baggage of the past included the discredited rhetoric of land and folk. Reconstruction meant re-education, the replacement of textbooks and also a change in the goals of education to a focus on economics, technology and the social sciences (Phillips, 1983; Trommer, 1990). Not only history textbooks, but works on biology, chemistry and landscape planning had to be rewritten (Groening and Wolschke-Bulmahn, 1987). Where old textbooks were reissued, as happened in the 'Emergency-textbook' programme, it was with a warning that they must be used with care. The 'biologic' point of view that saw man as one with nature had been part of the tradition encouraged by the Nazis. Understandably, then, biology received special attention; a branch of it, human biology, was held responsible for the Nazis' racial creed. The downgrading of biology, though, applied to all branches of the science, and was especially intended to purge the intensely ideological 'nature-loving' aspect of biological education that had been encouraged by the Nazis. Biology was to follow the methodology of an exact science, such as physics or chemistry, and take a low priority compared with them (Trommer, 1990, pp. 250–1). During the 1950s this rhetoric continued (ibid., p. 252). However, although the efforts put into re-education by the Allies succeeded in eradicating past ideas from textbooks, no clear alternative emerged, possibly because the same authors were writing them for the 1950s as had written them in the 1930s. The consequence of this dutiful rewriting by academics trained in a different and hostile methodology was that something had been taken out, but nothing very specific had been put back. A similar process, of forgetting but not relearning, can be seen in the case of history. Instead of history being taught as the Allies wished it to be, for many years in German schools history, especially recent history, was simply not taught at all. It was replaced by languages, classical studies, pedagogic and didactic studies, and work on comparative culture.

This 'blank page' approach to the past seems to have prevented the post-war generation developing the Manichaean vision of history which characterised later ecologists. The blanking out of history, the new beginning, also blanked out belief in the secret traditions of Green history, that mythical time of matriarchal and non-exploitative nurture, the alternative history of Paracelsus, Goethe and Spinoza that is such a powerful part of environmental thinking.

There were exceptions to this. One group that linked the present to a pre-war tradition was the anthroposophists inspired by Rudolph Steiner. The link between ecologists and followers of Steiner may seem obscure to the reader, and, indeed, it can be hard to find out from this secretive and non-evangelising group what they really believe. However, a reading of their early writings and Steiner's lectures on organic agriculture in the 1920s shows how closely their interests match those of ecologists. They talked of the chain of fertility, from sun to soil to man, resting on a subtle cosmological and magnetic balance. They preached harmony and self-sufficiency. Decentralised farms and villages, where man could live in close communion with nature as steward and guardian, were their chosen means. This organic farming tradition had continued from 1945 on, supported by the anthroposophists.

They published a journal, *Demeter*, and campaigned in a low-key (anthroposophists are opposed to political action) but persistent way against artificial fertilisers, factory farming and polluted air, soil and water. One minor example of this persistence, as well as of continuity, is that an attack by anthroposophists on I.G. Farben in 1938 for supposedly plotting to force chemical fertilisers on Germany was repeated in an article in *Der Spiegel* on anthroposophy in the 1980s. The original letter by I.G. Farben was no longer in the 1930s files, but it, or a copy, was apparently still in the possession of the anthroposophists.

Furthermore, despite the devastation and guilt of the immediate post-war period, there were still conservative thinkers who abhorred technology and industrialisation, and even, ironically, saw a chance of salvation in the economic ruin around them. Where Hans Morgenthau had proposed the pastoralisation of Germany, writers like Friedrich Georg Jünger wrote attacking continued development in technology, and hoped that Germany would remain 'undeveloped'. Wolfgang Siedler, a publisher and former friend of Albrecht Haushofer and his circle of conservative resisters, circulated material in 1964 which attacked proposals for reconstruction, and demanded the continuing pastoralisation – in the real sense – of Germany.*

Nonetheless, the political framework created by the need to reconstitute Germany did not allow for environmental concerns. Politics came after economics. The 1950s were dominated by controversy over how Germany was to be incorporated into Europe, how the new Germany was to be recognised. The Social Democratic Party supported a neutral and unified Germany, and Christian Democrats a Westernised Germany, at the cost of unity. This was not a real alternative, and although tensions arose when Germany was on the point of joining NATO, and the Soviet Union offered Germany unification in exchange for neutrality,

---

* Communication to the author.

there were no mass demonstrations or united opposition. There was some feeling against the establishment of a German army, and tentative anti-nuclear protests took place in the late 1950s and early 1960s, but the Cold War was too recent and too real to allow any serious dissent. The German student protest movement of the late 1960s was a fertile seedbed for later Green activists. German left opposition was well organised, but largely confined to the educated middle class, especially students, just the constituency who were later to support the Green Party. Despite the existence of several communist groups, Marxist-Leninists and Maoists were more prominent, and some Maoist groups were later to join in the first Green coalition, the Alternative List (student leader Rudi Dutschke was also to become a Green supporter). The revolutionary anarchists of 1968, inspired by the liberation rhetoric of the New Left, were the first political group in post-war Germany (apart from right-wing conservatives, an esoteric taste at the time) to attack bourgeois ethics and materialism. While the involvement of former members of this group – now suitably embourgeoised and prosperous – in the Green Party is one explanation for their broad left position on many issues and lack of issue-oriented environmental policy, another is the shadow of the past. German Greens were, and are, particularly worried about the pre-war link between Nazism and rural rhetoric and anxious to avoid being tainted with right-wing thinking of any kind.

This may be one reason why the call for ecologism by writers like Jünger did not attract support from the intelligentsia or the public. Clearly, in the 1940s, individualistic conservative calls for values could be attacked, and were, especially by the left, as reactionary and irrelevant nostalgia. However, Jünger's work extended beyond the 1946 work described above. From 1970 on he co-edited a journal of alternative ideas and in 1975 he published a long manifesto on the environmental situation which dwelt on ecological themes. It aroused little interest, and does not seem to have become part of the Green package-deal. Certainly, it is not mentioned in books about German Greens, or works by German Greens themselves. Why should this be? The manifesto itself seems squarely in the tradition of ecologist writing. It talks of the natural cycle of fertility, of natural resources, and the danger that pollution and poisoning of resources within this cycle will result in catastrophe. It attacks industry and prophesies environmental disaster as a result of industrial side-effects, but also because of the loss of biological species, their richness, diversity, colours and textures. Humanity needs the natural world if it is to remain fully human.

In a sense, right ecologism had come too late for its anti-industrial and aesthetic critique to influence public opinion. Jünger's *Bussauer Manifesto* appeared three years after the Club of Rome's *Limits to Growth* (Meadows *et al.*, 1972), which was enormously successful in Germany, selling up to 500,000 copies. Another reason why this conservative-oriented document fell on deaf ears may be that, as Mohler points out, Jünger's approach was *defeatist* (Mohler, 1978). The same could be said of Herbert Gruhl's *The Plundered Planet*. Indeed, this darkness of vision is perhaps why the dissident conservatives' approach to ecologism was overtaken by the left-liberal German Greens. Despite the general apparent pessimism of the ecological analysis, the emphasis on catastrophe, apocalypse and the Doomsday syndrome, its corollary is the optimistic belief that, given sensible education and a reasonable approach, people will listen and governments will change their ways; that reform is possible. Participation and education are left-Green nostrums. They relieve the blackness. Germans who were single-issue ecologists stemming from a conservationist and nature-loving tradition suffered from 'cultural pessimism'.

And, the party-oriented left-Greens were to be proved right, though for unexpected reasons. Not through democratic participation, decentralisation and education, but eventually through their political success, they would force Green reformism on to the agenda of the main political parties. And if today's staunchest Greens are to be found among the Bavarian farmers who support the CSU (Christian Social Union), that too is consonant with the early days of the German Greens, who began their political existence in association with a group of farmers from Schleswig-Holstein.

Britain 1945 to 1970

In *The Global Environmental Movement*, John McCormick has argued that the period between 1962 and 1970 saw the transformation of the environmental movement into a powerful force. This was especially so in the USA, a country he sees as leading organised environmentalism (McCormick, 1989). It is true that Britain to a large extent followed America in the growth of organised lobbies and mass mobilisation on environmental matters, so much so that some commentators see it as an American phenomenon. Yet Britain has had a scientific ecological movement since the turn of the century. The organic farming movement spawned several pressure groups, while by the mid-1960s the term 'ecology' had become familiar. Britain's home-grown environmental movement, dating back many decades, had already expressed itself, however inadequately and incompletely, in the concept of the 'green belt', the town and country planning legislation that affected where and what could be built nationwide, the Clean Air Act (the world's first), and the attempt to decentralise offices and government departments.

This list of political gestures towards the ideal of preserving the countryside, controlling development and shifting the population out of big cities could be extended. They grew out of the radical Liberal, garden city movement of the late nineteenth century, and fitted into the post-war consensus of local and national government intervention and planning. The fact that garden city ideals were incorporated into 'normal' politics meant in turn the exclusion of environmental projects that did not fit in with the prevailing ethos.

By 1960 nothing could have seemed more irrelevant to a diagnosis of Britain's ills than environmental politics. Whereas the United States had experienced almost uninterrupted affluence since the Second World War, Britain had seen rationing, shortages, and an increasingly impoverished middle class. She was especially dogged by policies evolved before the war, which no longer related to her role as a post-imperial power with a mobile population, in economic decline. This meant that the 1960s saw the climax of the concept of growth and technology, enshrined in the unfortunate phrase, 'the white heat of the technological revolution', with which Harold Wilson, perceiving the need for modernisation, ushered in an era of public growth and private stagnation. The 1945 Labour revolution had introduced a whole class of policy-makers and civil servants who saw their role as to destroy the past, and move into a Wellsian future. The 1960s saw their triumph, not through the infrastructural investment needed for roads, rail and sewers, but through a fantastically wasteful programme of city centre redevelopment and public housing programmes.

In the what-is-wrong-with-Britain range of complaints pollution came low on the list. Britain was backward, full of out-of-date machinery, being left behind. Few voices opposed this view. There was a tendency to interpret attempts to conserve the countryside as the selfish response of a threatened landowner class. I have argued that environmentalism had to move left to become acceptable to the media and to grant-givers who help to form public opinion and whose support is essential in forming the seedbed of new political movements. There was media prejudice against any social criticism coming from the 'right', as can be seen by the relative treatment given to two books criticising economic growth that appeared between 1967 and 1973. The economist E.J. Mishan wrote in 1967 and again in 1969 about the evils of growth, but his work received little attention (Mishan, 1967, 1969). Now that the idea of environmental economics has become fashionable, these books are belatedly being reissued. Mishan's work used economic theory and demanded a thinking reader. His unashamed appeal to values was exactly the kind of elitist idea that was jeered at at the time. The contrast with the fate of *Small is Beautiful* (Schumacher, 1971), which became a cult book all over the world, could hardly have been greater. Schumacher's work had that 'why oh why' quality of communitarian preaching that has mass appeal. But it also had a gut anti-Western attitude that appealed to the opinion-forming classes, the growing global community of problem-finders, who had occasional doubts about the validity of their path.

Schumacher himself had been an influential policy-maker and government adviser for some decades, something that perhaps helped form his preference for an authentic and intimate way of life. An *émigré* from Germany to Britain just before the Second World War, from 1945 on he helped to create the post-war welfare state; was economic adviser to the British Control Commission in Germany and the National Coal Board in Britain, and also advised the government of Burma after the country became independent from Britain and adopted an autarchic socialism. In 1962 he advised the government of India on rural problems, and was struck by the unsuitability of Western aid and technology to that country.

His career as lobbyist showed the importance of non-party organisations in propounding environmental ideas in Britain. He founded the Intermediate Technology Development Group in 1965, and became president of the Soil Association in 1971. He lectured in the United States, where he was enormously popular with students. As a German refugee, Schumacher brought with him the hostility to free market mechanisms and capitalist values that had characterised German intellectuals of most political persuasions for a century, and the moral infusion of thinkers in a country where the pastor was one of the strongest pedagogic and social influences. After all, even 'social market' theorists like Ludwig Erhard, creator of the German economic miracle after the war, began from a Christian and communitarian position, which concentrated on the duties rather than the rights of the individual.

But Schumacher also brought with him that romanticism about 'other', Eastern values which characterised German ecologism. In one of his more striking passages he alleged that the 'pressure and strain of living in say Burma' was less than that of living in the United States, despite its labour-saving machinery. (I owe this reference to Virginia Postrel, editor of *Reason*.) Schumacher was, after all, fairly familiar with life in Burma, if only with life as lived by government advisers, and one might have expected evidence to substantiate this allegation. Controlled economies can of themselves be remarkably stressful. The rate of heart disease has gone up or stayed up in most Eastern bloc countries since the war, and increasingly so since the 1970s. As Burma had been a closed, socialist society for some years when this remark was published, the comparison between Burma and the USA does not speak highly for his sense of realism, his authentic and intimate experience of life as lived by the masses in Burma. The word 'say' in the sentence quoted above implies a certain vagueness, that any old foreign country would do, especially in the romantic East; but the very vagueness of the approach to non-Western countries suggests that Schumacher is idealising an 'Other' world simply because he needs to find a contrast to the bad West, rather than relying on empirical evidence that would have been within his grasp had he chosen to look for it.

There was a similar gap between ideal and reality in the fact that he advised the National Coal Board (NCB). British coal is high in sulphur; it is dirty coal. Schumacher cared about environmental pollution, yet he continued to advise a body which was dedicated to the continued use of the dirtiest hard coal in Europe; which had to keep opening deep mines in order to produce a dirty coal that the electricity-generating system (also nationalised) was forced to use for its power stations, at the cost of many thousands a year who died or were made very ill as a consequence of the pollution involved.

Schumacher may have spent years in private suggesting to the NCB that they close down so that Britain could go over to natural gas and imported cleaner coal. I do not know. But there is no trace in his published writings of any awareness that he was involved with a major – and unnecessary – polluter, which was supported purely for political reasons. (The first post-war Labour government could not have afforded to import coal; the second needed the miners' vote. By 1972 the political power of the miners was such that their strike helped to bring down the second Conservative government of Edward Heath, while over the entire period Britain's need to export and cut imports meant going for self-sufficiency where possible.) Schumacher, then, despite being an economist, was not just an economist putting forward environmental arguments. His impact was through his appeal as a moral leader enunciating a creed – that every individual should think about the environmental implications of their lifestyle and restrict unnecessary consumption and ambition, so that the ethic of untrammelled economic growth could begin to lose support. There seems to be evidence that towards the end of what appears to have been an enormously successful life preaching the gospel of *Small is Beautiful*, Schumacher began to doubt whether his views would prevail (Davy, 1984). He seems to have thought that his phrase, which was now part of the language, had passed in too easily, that commitment to growth and big business was too entrenched to be easily changed, that a complete change in the West's world-view was needed and, failing such a change, that Western society was bound to destroy itself within decades.

The second guru to argue that the West was developing in the wrong direction was John Papworth, a pacifist and former editor of the pacifist, anarchist journal, *Resurgence* (which is now the voice of the alternative education movement, and published by associates of the Schumacher Society). Much of the thrust of Papworth's criticism was directed at Western domination of the Third World. The novel aspect of his attack was that he criticised the whole idea of modernising the developing countries by encouraging urban growth at the expense of rural life, of introducing advanced technology and large-scale machinery to countries which needed better wells, small dams and motors for fishing boats. Once this criticism of wasteful and destructive development had been expressed convincingly as far as the developing world was concerned, its implications could be extended to the developed one.

As with the Sierra Club in America, the history of the political and cultural shift of environmentalism can be demonstrated in the history of one organisation. The Soil Association was as far from the spirit of the National Coal Board as can be imagined, yet E.F. Schumacher was its chairman from 1971. It was created from the impetus of Lady Eve Balfour's influential book, *The Living Soil* (1941). This work attacked the use of pesticides and other chemicals in farming, for the same reasons as those later given by Rachel Carson in *Silent Spring* (1962), though without the supporting evidence of a silenced earth. Balfour wrote of a world covered by a living organism, the soil, humus rich, created by micro-organisms, worms and naturally composted mulches. Man depended on the chain of sun–soil– food, she wrote, and should not introduce dangerous chemicals into it, but should imitate nature's method of soil maintenance. Lady Balfour went on to start experimental farms, and founded the Soil Association in 1945. Several landowners who practised organic farming on their own

estates were on the board, including three who had had
strong right-wing sympathies before the war (Bramwell,
1989), but the association was non-political. Its strongest
influence came from Anthroposophy, and many members
believed in vitalism and holistic science. Others supported
land redistribution, and wrote glowingly of Chinese communes. In so far as there was a political ethic, it was that of
stewardship, with the mildly feudal overtones that implies.
Values, and the need for individuals to fight for those
values and live according to them, were stressed. Most Soil
Association members also believed in alternative medicine,
communes and crafts; many were vegetarians. The dearth
of mildly feudal and charismatic leadership was such that in
1969 it was Barry Commoner, on the 'left' side of the
population debate in America, who became vice-president.

The philosophical side of environmentalism was, in
the meantime, propounded by Edward Goldsmith, editor
of *Ecologist*. His *Blueprint for Survival* (1972) was a crucial
text for restructuring socio-economic life: it argued that
economic growth must be reversed, population growth
stopped and then reversed, in order to avoid a catastrophic
overuse of resources. Indeed, predictions of apocalypse
were a significant milestone in the development of environmentalism in the 1960s. Writers today sometimes think
that it was the oil price rise that produced resource fears,
but these pre-date that shock. However, it is true that
during the 1960s panic was fuelled by population growth,
especially though sometimes implicitly in the Third World,
rather than the finitude of resources.

In Britain, the Campaign for Nuclear Disarmament and
other pacifist causes were powerful in the 1960s. Student
radicalism only took off in the late 1960s, and did not at
that stage have an environmental component. The anti-
nuclear movement and the power-to-the-streets anarchist
movement provided activists who would be prominent in
Green politics a decade later. The Green movement in
Britain is seen as more left than in other countries,

Germany excepted, and this may be because of the void caused by the decline of organised socialism in this country, a decline not unconnected with the twelve years of Labour rule between 1964 and 1979.

It is interesting that left disaffection increased during the 1960s, for there was a six-year Labour government in office during this time – after a fifteen-year period of Conservative rule. Radical socialism was not the keynote of the Wilson government, which wanted to prove itself fit to rule and after a brief economic crisis went for sound money, economic growth and efficiency. It remained in NATO, kept the nuclear deterrent, continued subservient to the United States regarding defence policies, and tried to enter the European Community.

If there is a link between economic growth and interest in environmental affairs – and I referred earlier to the suggestion that post-materialist values emerge following a sense of economic satiety – then it is hardly surprising that Britain, dogged by economic problems, low pay, chronic inflation and low mobility, should have failed to develop a mass movement concerning environmental matters during this time. In any case, as long as environmentalism was perceived as a middle-class-oriented movement, it was unlikely to succeed.

CHAPTER THREE

The USA 1970 to the Present

ENVIRONMENTALISM IN THE USA since 1970 has seen the intertwining of two very different strands: New-Age-style introverted ecologism, and the influence of former Marxists.

At first sight it may seem that the reason why no Green party was formed in the USA is that in a two-party system it would have little chance of success. However, that did not prevent the Green Party being formed in Britain. More probably, it was the surprising success of the environmental lobby in the USA that vitiated the need for party political activity. A variety of single issues had been fought over and won, largely thanks to Washington lobbying. Nonetheless, party activity offers something to political activists, especially those with a statist leaning, that they cannot find in single-issue pressure groups. It offers continuity, hierarchy, authority, cadres. If the victory of President Clinton's Democrats had not filled the gap, there would by now be a Green party in the USA. The presence of environmentalists in Jesse Jackson's Rainbow Coalition, and the range of ideologies involved, parallels the early period of the German Green Party, but because of the strength of the American environmentalist lobby, a variety of alternative causes, including civil rights, pacifist and anti-nuclear struggles, will battle for supremacy with environmentalist issues. Indeed, battle lines were drawn up at the Green Convention of 1987, at Amherst. Here left-Greens argued that capitalism was the enemy, while deep ecologists saw intrinsic human nature as responsible, and warned that systems should not be made the scapegoat for the problem.

The political split behind the environmental movement – essentially New Left versus New Age – dominates other divisions, such as that between radical, direct action and legal action, or deep versus reform ecologists. It affects the self-image of environmentalists in America, who are able to use local issues to score points with some success, but who feel powerless against the might of oppressive big business and organised capitalism. Yet although they perceive the situation regarding pollution as constantly worsening, the USA is nonetheless the country where environmentalist actions have been *tactically* the most successful. How does one judge success in this context?

Earlier I indicated the surge of environmental legislation during the 1960s, culminating in the National Environmental Policy Act of 1969, and commented on the large amount of land owned by the American state – 42 per cent. By 1992 the Environmental Protection Agency, with some 18,000 employees, was the largest civilian agency in the federal government, and is expected to be given cabinet status under the Clinton administration. In California, the state authorities have set hard and fast goals for a percentage reduction in air pollution, of a nature that will force car manufacturers to begin to convert to alternative fuels from 1998. Despite the impression given by American environmentalists, this is quite an 'achievement' in a country with no Green party, and it has been due to the remarkably successful lobbying tactics of environmentalists since the early 1960s.

But because American clean-up legislation has lacked a cost-benefit approach, and because of the emphasis on toxic waste and soil, results were few. The money reclaimed from polluters or their heirs was supposed to be ploughed back into clean-up, but most of the resources went on court cases. Soil pollution has deep psychological significance, but is not the worst pollution problem in terms of damage either to human health or to the natural world. Until the recent experiment in trading air-pollution permits, American environmental legislation also assumed, as hitherto most American environmentalists have, that private enterprise and capitalism is the problem, because exploitation, greed and the profit motive cause environmental damage, and that federal regulation is the only means to control otherwise uncontrollable and malevolent businessmen. The weakening in the lobby structure under James Watt, Reagan's environmentalist supremo, began to produce, ironically, a shift of tactics among environmentalists, who started to support the buying up of wildernesses (and other areas threatened with development) by private trusts such as the Audubon Society.

So while groups like the Sierra Club are delighted at the increase in their political power, many members have become disillusioned by the failure of an increase in legislative and other more direct forms of control to have the desired effect. Of course, some argue that more and more effective control is needed. Although studies of business behaviour suggest that American companies are worried about environmental hazard assessments and try to comply with them (fearing massive lawsuits if they do not), and that American-based multinationals have forced their subsidiaries in Europe to abide by higher environmental standards, the perception among environmental activists is that the legislation is virtually unenforced. Because pollution incidents often date back to before the enactment of environmental legislation, forcing companies to accept responsibility for past hazards and correct them has proved arduous and frustrating. The attempt to produce a low-risk environment has also attracted protest. Some writers have accused environmentalists of exaggerating the dangers from pollution, and of seeking an impossible, risk-free utopia (Kahn and Simon, 1984). But the application of orthodox economic logic would claim that risk prevention in crowded urban areas need to be based on stricter standards than, for example, in the

Nevada desert. This method of arguing is alien to environmentalists, who point out, with some justice, that the wilderness, with its potential impact on the food chain, air pollution and animal life, needs protection as much as human beings living in a metropolis.

The pro-regulatory stance of the environmentalist movement goes back to the anti-Vietnam movement of the late 1960s, many of whose veterans appeared to be involved in a complex game with the authorities, which ended with their 'long march through the institutions'. A surprising number of today's environmentalist activists were active in the student protest cause, and then shifted interest rapidly to environmental causes in the early 1970s. Since environmentalists' values are not especially left-wing, nor their strategy inherently so, this development is at first sight puzzling. It may be due to disillusionment with the orthodox left, coupled with a desire for a new cause among skilled activists.

In *The Sand County Almanac* in 1949 Aldo Leopold claimed that the sturdy independence behind American democracy needed the wilderness experience to reinforce it. Although today this comment could easily pass for an extreme right-wing assertion (because of its implicit emphasis on the white American heritage), the backwoodsman ideology in the pristine form of the frontier experience, despite its sturdy individualism, is not especially conservative, but is almost anarchic – oriented towards a liberal rather than an authoritarian stance. This anti-authority approach easily fits into environmentalist ideology, as does the overriding priority of conscience over duty to established authority. Early American environmentalists included liberals like David Starr Jordan, and Joseph LeComte, an earth sciences specialist who wrote in the American *Monist* along with phrenologists. Both were eugenicists. In the United States, rights-based ideologies turn naturally to legal activism.

Marxist and socialist activists adapted some of Marx's criticisms of capitalism where they matched the ecologists'

critique. Reification – the process whereby a worker becomes alienated from the product of his labour – had a
twofold explanatory structure in Marx's work. The technical structure of Marx's argument, based on his politico-
economic model of the dual commodity quality of labour,
was difficult to grasp, and lacked gut appeal. But the emotional argument – the effect of the division of labour on
worker attitudes; the separation of worker and object –
could be more easily understood. The environmental damage caused by capitalist agriculture had been attacked by
Marx (although he had no time for agricultural labourers,
rural beauty or peasants). The emotional attacks on the
profit motive, on the greed of capitalists, are echoed by
heartfelt attacks by ecologists on greed and materialism as
agents of destruction.

The rise of the guru

The ideology behind the Greens in America, and hence,
through the transmission of a dominant North American
culture and politics, behind Greens in Europe, has been
warped away from a pure environmentalism by the late
1960s social criticism of Marcusians and disaffected Marxists. As discussed earlier, the hunt for values, the preference
for authenticity, meaning and tradition, the attack on a
centralised, anomie-producing state are major themes in
Marcuse's works, and parallel the cultural criticism of capitalist society to be found in the work of nineteenth-century
German and British conservatives. The distaste for industry
and the resentment of technology are common to both.
'Left' and 'right' are problematical terms here. How does
one place Ruskin, Matthew Arnold or even William Morris,
who despite his support for the workers' revolution expressed nationalist views that would not be acceptable to
the Labour Party today (Bramwell, 1989)? There are con-

tinuing disputes about whether their support for cultural and aesthetic values is 'left' or 'right' oriented. Anyone calling for social reform through changes in *individual* consciousness and behaviour belongs to one end of an individual–institutional axis, while anyone preferring to work through state institutions belongs to the other. In the discussion on the German Greens I pointed out that, apparently paradoxically, the *Fundis* (opposed to compromise with the state and system) have alienated those Greens who were formerly Marxists and who have a gut sympathy, even a love–hate relationship, with institutions. Similarly, one force behind the development of American environmentalism was the cultural and moral criticism of the manipulative, powerful and technology-backed state expressed by the gurus (of whom Schumacher was one): often using the radical jargon of alienation, calling for decentralisation and popular power, their actual prescriptions are usually vague or even comical. That does not lessen the force of their impact. On the contrary, they appeal to the tradition of preaching, of claiming the moral high ground, which is to be found in politicians in the USA, Britain and Germany.

Two seminal 'guru' books appeared in the USA in 1970 and 1971 that both continued the development of the new ethic outlined above and led it. Both were published in Great Britain in 1971. They mark the apogee of the transition between the New Left movements of the 1960s and the New Age environmentalism of the 1970s and 1980s.

Ivan Illich's *Deschooling Society* is an important example of the reforming ecological ideal. Although his book *Medical Nemesis* was also influential, it appeared some years later. Nothing made quite the impact of his first published work. Why do I put him with ecologists, when he is known as a controversial quasi-anarchist educational theorist? For two reasons. The first is that his book is saturated with comments about the coming global catastrophe caused by pollution: his attitude to technology is ecological. In short, he wants technology to work for people rather than to impose systems and structures upon them. Just as the communal ideal was part of the restructuring of society, its resource use, family habits and sexual relations, so a new kind of education is a concept crucial to the New Age vision. In Britain, Illich's vision helped to produce the Small Schools movement, supported by names like Schumacher, Papworth and Kumar. In America it paralleled libertarian criticisms of the educational system, and the apparent fruitlessness of pouring more money into it. Secondly, Illich corresponds to the political profile of the New Age ecologist. In his writing he defines himself as 'left', and categorises as right those institutions that *process* people and their lives, but his underlying aim seems to be a call for more popular control as against manipulative institutions and technologies.

In the discussion that follows, some quotations that were in an earlier draft of this book, which revealed the more eccentric or inconsistent sides of Illich, have been deleted, because there is a danger that the reader's natural reaction would be to ask why, if Illich's works included unreasonable and unsustainable arguments, he should be treated with such reverence, and why he caused such a stir. One answer is that Illich produced criticisms of contemporary society that could easily be packaged for mass and uncritical discussion in the media, with their dottier aspects taken out. Second, his erratic and potentially unpopular claim – that all virtue resides in the 'left' and all evil in the 'right' – could be argued and reinterpreted, rather as Rousseau is constantly being reinterpreted, and as failed communists in the former Soviet empire can be described as 'conservative'. What Illich really meant by the system he attacked, it was argued, state socialism as well as state capitalism. The term 'left' included libertarian, liberal, small-scale, intimate social arrangements. 'Right' meant 'them': advertising executives, military men, weapons manufacturers, big business, but it did not refer to individuals, only to institutional structures. The guilt lay in the system, not in the person.

Third, Illich's points offered the disaffected or powerless left-intellectuals in the USA a way out. They could adopt a new ethic, admit that state reformism and interventionism over the last fifty years had failed, without having to join their traditional enemies on the right, those reactionaries who had always opposed the extension of state power and state institutions. Now there was a new state concept: the benevolent, non-interfering state, which laid down a framework to protect and encourage popular 'autonomy', which would help but not force people to be free. Many of those who had worked in state institutions all their lives, particularly those involved with global policy-making, were disillusioned with the theory that Western development patterns should be implemented in developing countries. Illich's realisation of their worries was manna.

It is in any case the book with the profound emotional charge that will express the anguish of a time, not the arid rationalisms of commentators who wish to normalise a critique they find attractive.

He begins by analysing a state in which 'right' and 'left' have 'service institutions' (Illich, 1971, p. 60). With what Illich calls 'right-wing institutions', that is manipulative, modern and alienating institutions, 'the client is made the victim of advertising, aggression, indoctrination, imprisonment or electro-shock'. It is an example of Illich's unworldly and moral fervour, his interest in the state of humanity's soul, that he should seem to give equal weight to the horrors of advertising and those of electro-shock. All institutions force a consumer addiction on the client, either by advertising or by compulsion. They are 'socially or psychologically "addictive"' (ibid.): consumers come to desire more consumption because of the artificial stimulation of dissatisfaction with what they have already consumed. By contrast, he sees 'service networks' of the left as self-limiting, and not addictive. One example is the telephone. You use the phone to make a communication, not just to talk into the receiver. Although it is an advanced technological product, it is also a convivial experience.

Illich's use of the term 'right-wing' institution and his idea of a left–right spectrum are curious, and even clumsy, because they do not convey his overall argument. Although he focuses on the systems established in the West, that is under capitalism, the institutions, agencies, schools and corporations, his attack on a system where people do not feel in control of their own lives, his opposition to manipulative institutions, places him in opposition to centralised state systems of any kind, whether capitalist or socialist, although with an emphasis on hostility to technologically advanced societies. It is the manipulation from above that he resents, whatever the label. He happens to put small crafts (hand laundries, small bakeries, music teachers) at the left end of the spectrum (and hotel chains in the centre), but to a political scientist or historian the world of the small shopkeeper and the artisan does not correspond with established definitions of the 'left'; rather the contrary. Small shopkeepers and artisans have traditionally supported Poujadist and early Fascist movements.

A significantly ecological argument lies behind Illich's attack on cars. According to Illich, cars are right-wing goods; they exist merely to service the money men of General Motors, while roads exist merely to service the cars (ibid., pp. 62–3). Access to roads is confined to those with cars; a road is not a genuine public utility, like a telephone network. Life in a car is not convivial, unlike life in a bus. Illich is not arguing against mobility in itself, but is anxious to establish a resource-effective programme for developing countries so that they can avoid the West's pattern of personal consumption. He thinks that spider-web networks of trails should be built which enable all regions to have slow, reliable, utility transport, low maintenance buses, in effect. Illich comments that a demand for such vehicles would 'have to be cultivated, quite possibly under the protection of strict legislation'. This sudden call for state control in the middle of the anti-statist appeal to individual values is again typical of the ecologist, as is the assumption that something has to be done for developing countries, and obviously must be done by the West, to enable these developing countries to fulfil their own, different destiny. The idea that they should actually be left alone to do so is alien to him, as is the idea that if human beings desire individual transport, there may be some virtue in it.

Illich is known best for his approach to education, which again, but apparently unrealised by him, strikingly cuts across political boundaries. He sees schools as artificial, constraining, ineffective in achieving true education, time-wasting, money-wasting instruments of inequality, negative artefacts. His alternative is to give the population education credits, to be spent at any time, while education will be acquired as it is already, through experience at the workplace, talking with colleagues, learning skills. Skills can be learnt more quickly and flexibly; ideas can be exchanged, access to laboratories, factories, farms and libraries will teach how things work, and, most controversially, children should work at least two hours a day.

Illich is prepared to use technology if it can lead to the kind of society he wants to see, especially if it can be used to convey information and prevent the waste of resources. However, he does not see that ownership and the protection of ownership have any relationship to the functioning of a society. In his educational reforms he dodged the problem of how knowledge was to be transmitted free of cost. This led to the situation where, for example, he supported the setting up of a computer network that would match up the interests of those seeking discussion to enlarge their knowledge. He also supported skill exchanges, peer-matching, educationalists at large and educational games. He was even prepared to pay companies for letting people use them as 'learning facilities'. He would abolish certification and testing.

Illich's demand that traffic should be banned from New York so that adults and children could visit 'storefront depots' of knowledge without being run over combines an advanced view of inner-city traffic control with a somewhat narrow focus of demand (ibid., pp. 85–6); an example of the 'why don't they do this' approach which characterises the powerless.

Illich's point that education is a means of socialising children, rather than of truly educating them, that the money poured into education, especially remedial education, is ineffectual, and that Third World countries have wasted resources and increased inequality by using compulsory state education as a means to modernisation was a brave criticism of established dogma. It saw through the wishful thinking that had grown up about 'manipulative systems'. Yet his reform proposals seemed fuzzy at the edges, based more on wishful thinking than on the kind of perceptive criticism with which he began his progress.

He begins his most famous book, *Deschooling Society*, by linking pollution with the schooling process.

> the institutionalization of values leads inevitably to physical pollution, social polarization and psychological impotence: three dimensions in a process of global degradation and modernized misery.

Illich's comments on global pollution are especially interesting because he takes it for granted that his readers will understand and agree with his criticism:

> Modern agriculture poisons and exhausts the soil. The 'green revolution' can, by means of new seeds, triple the output of an acre – but only with an even greater proportional increase of fertilizers, insecticides, water and power. Manufacturing of these, as of all other goods, pollutes the oceans and the atmosphere and degrades irreplaceable resources. If combustion continues to increase at present

rates, we will soon consume the oxygen of the atmosphere faster than it can be replaced. (Illich, 1971, pp. 110–11)

And,

> It is now generally accepted that the physical environment will soon be destroyed by biochemical pollution unless we reverse current trends in the production of physical goods.

He continues, tying up this criticism with his basic opposition to schools:

> It should also be recognized that social and personal life is threatened equally by HEW [The US Health, Education and Welfare Department] pollution, the inevitable by-product of obligatory and competitive consumption of welfare. (Ibid., p. 17)

Illich's approach is from the New Left, but as with other New Left attitudes, the Old Right is not far behind. Surprisingly, the Old Right too attacks consumption. Capitalism, advertising and fast cars throw them into hysterics. Comte Gobineau and Joseph De Maistre, the French extreme conservatives, both thought that compulsory state education would diminish human competence and creativity. However, they would not have shared Illich's preference for the undeveloped world, although both saw the spread of Western culture and technology as pernicious to the recipients. Where the conservative sees technology and modernity as *too* liberating, as sacrificing true social values for a consumer satisfaction that is only apparent, Illich sees them as constricting, because impenetrable to the understanding of the consumer. Modern cars are worse than old ones because no one can repair such cars themselves. Transistor radios are worse than old radios because they are incomprehensible to the ordinary consumer. Consumer

satisfaction is in general spurned by Illich. He has no understanding at all of why anyone should want to travel in a car,
or have a choice of goods instead of one brand only; and he
sees capitalist competition as absolutely wasteful. To him, it
serves no corrective purpose at all. The libertarian criticism
of welfare and compulsory schooling (education credits are
now a standard part of their programme) came to similar
conclusions.

Revealing both his theological background and the religious quality of environmentalism, Illich argues that hope
is better than expectation:

> Hope, in its strong sense, means trusting faith in the good-
> ness of nature, while expectation, as I will use it here,
> means reliance on results which are planned and controlled
> by man.

Despite the Enlightenment quality of some Green thinking,
we have here a desire to escape from contingency, from
cause and effect, from effort and reward, that is entirely
utopian.

Illich is like other ecological gurus in his attitude that
difficult problems in his reform have still to be solved, but
he is sure they can be. As a theological student, then
educational and medical specialist, and assistant rector of a
Catholic university, he fits the pattern of the *Beamten* or
middle-class, professional civil servant ecologist: reacting
against the state that supports him, aware of the deficiencies and creative death of the 'from above' system, but still
hostile to what many would see as an alternative autono-
mous world of capitalism, where consumption is in the
hands of the individual. Illich did not differentiate between
coercive institutions (such as schools) and institutions such
as a shop where the consumer has some freedom of choice
if not a perfect freedom.

Nietzsche too attacked a version of 'learning', in his essay
on 'The Uses and Abuses of History for Life'. He saw it as a

process destructive of cultural creation. Knowledge of the past, of past myths, religions and achievements, would always tend to hamper free creative thinking in the present. Some evidence in support of this thesis appeared with Charles Reich's seminal work, *The Greening of America*.

This work was first published in 1970 and again rapidly republished in Britain. Despite its title, the Greening in question is not specifically oriented towards ecologism. However, issues of pollution, environmental degradation as well as generalised cultural criticism of modern urban society are mentioned constantly, and form part of Reich's overall picture.

Reich was writing at the peak of what we now call the 'sixties', which seem to have started in 1968. Some of his comments strike the reader oddly now: his hymn of praise to bell-bottom jeans, for example. However, at the time Reich expressed the alienation from corporate America that dominated the white middle-class, student body. Now that that middle class is seeing the fruits of its long march through the institutions, the form of this alienation is relevant.

Reich was a forty-two-year-old lawyer at Yale University when the book was published, and the concentration on a new, young generation which embodied a new consciousness – Consciousness III, he calls it – is striking. Also striking is the Marcusian and hence small 'c' conservative nature of his cultural criticism. He is against anomie, alienation, big cities, lonely apartment blocks, noise, technology and change. Work is artificial; democracy is a fraud, a system suited to small-scale societies helpless in the face of organised big business and overarching bureaucracy. Society is commercial, materialistic.

Technology and production can be great benefactors of man, but they are mindless instruments: if undirected they roll along with a momentum of their own. In our country they pulverize everything in their path: the landscape, the natural environment, history and tradition, the amenities and civilities, the privacy and spaciousness of life, beauty and the fragile, slow-growing social structures which bind us together. (Reich, 1971, p. 4)

Reich argues that meritocracy, part of Consciousness II, breaks up communities, fragments families and draws the able from their background (ibid., p. 6). This argument in favour of a static elite, a hereditary hierarchy, could have come from Victorian England, *circa* 1838, or the Germany of 1890, except that Reich saw salvation in the Eldridge Cleavers and the rock music of the new consciousness.

Reich emphasised approaching doom, caused by the collapse of the corporate state, the meritocracy, the hard, unloving, unplayful pattern of work. He paints an alarming picture:

For most Americans, work is boring, mindless, hateful . . . America is one vast, terrifying anti-community . . . America is dealing death, not only to people in other lands, but to its own people . . . disintegration of the social fabric . . . atmosphere of anxiety and terror . . . lawlessness and corruption . . . drastic poverty amid affluence . . . de-personalization, meaningless repression . . . loss of self, or death in life. (Ibid.)

He sees the answer in the new young generation, the college, campus hippie. They can live with our technology, they can take it or leave it. They have adjusted their lives. The young have already evolved the answer; they are saved. From death comes regeneration 'in this moment of utmost sterility, darkest night and extremest peril' (ibid., pp. 13–14).

This emphasis on death and regeneration, common among ecologists, is the language of the religious visionary. Reich sees the saved everywhere on his campus. He believes that society will spontaneously fragment and reconstitute itself in the form of communes. But 'ultimately, as the film *Alice's Restaurant* clearly shows, intimate communities will have to be based on something more than love or a common "trip". The most successful communes today seem to be the rock groups that live and work together' (ibid., p. 284). Reich is keen on the collective, holding to the naive belief, deriving from Engels' historiography, that before the Industrial Revolution the extended family was the norm, man lived close to his work, was not alienated from his labour, and so forth. He tells us, in the tradition of collectivist aesthetics, that 'Art and workmanship came out of a cultural tradition, not an act of individual genius' (p. 285). Consciousness III is to restore this medieval wholeness, this oneness, this loss of ego, of selfishness: 'It is no accident that marijuana joints are always passed around from hand to hand and mouth to mouth' (ibid.).

The social norm he sees emerging has, significantly enough, the college dining hall as its model. These are, it seems, uncompetitive, sharing, informally happy places. They buzz with intellectual life, talk and jokes. Under the Consciousness III lifestyle, there is more: unguarded smiles, open gestures: '*it is the atmosphere of breaking bread together, of communion*' (p. 286; my italics). Reich's crescendo is a description of a supermarket in Berkeley (near the campus of the University of California). A modern Berkeley supermarket is indeed a marvel, and the mix of people in it is sweet and enlivening. Reich does not analyse the market forces that make it possible, the prosperity that brings 'a veritable one world of foods' and makes these available to the poor as well as the rich; the sheer wealth of a country that can offer its youth education for almost a third of their lifespan. He only sees a community, a oneness. 'Somehow all these people were together. . . . The scene as a whole . . . was not a march but a kingdom – the peaceable kingdom of those old American paintings that show all manner of beasts lying down together in harmony and love' (p. 287). This again is a religious image, drawn from the well-known description in Revelation that has inspired so many paintings.

The glorious absurdity of Reich's vision, the pathos of this humourless law professor yearning for bell-bottom jeans, is quite beside the point. Reich was believed. He truly expressed the feelings of the fashionable, artistic and literary intelligentsia of the time. And the students of that era went on to take their unformulated cultural criticism, their religious yearnings, their belief in apocalypse and the saving of the few, into the ecological movement. Only a college-educated body, thoroughly alienated from their past, and including a substantial group of 'red diaper babies', as they were called, could have believed so uncritically. Not surprisingly, they were to believe equally uncritically in the conspiracy theory of environmental disaster and crisis in each successive decade in which it was formulated. And not surprisingly, the involvement of students and college faculty, the ideal based on student dormitories, dining halls and communes, remains the basis for the ecological utopia. This was a happy time for those involved, because they went on to good and safe jobs in the system they so despised. Just as the English public school played a role in the lives of its alumni hard to believe or understand by those who did not attend one, so the open smiles and unguarded gestures we are told dominated Consciousness III remained an ideal for those involved. And the fact that this ideal rested on vast tax expenditure, on families willing to subsidise their lifestyle, and on the introduction of a network of drug-dealing and rock music ancillary trades could be overlooked in the general belief in rights and claims against a faceless, corporate state.

Rereading Reich after a gap of twenty years, I found it interesting that no evidence is cited from start to finish. No data or analyses of crime rates, suicides, or standards of living are offered, but this did not diminish the book's success or lessen its impact. Despite its learned use of Marcusian terms, its talk of repression and consciousness, it

is personal and anecdotal in the extreme. But the fundamental message, that the corporate industrial machine was
about to self-destruct, reflected the spirit of the age so
closely that it was accepted and discussed uncritically – that
is, the emotional assumptions went unquestioned.

The active, lobbying environmentalists, in America and
elsewhere, are more sophisticated than the wide-eyed com-
munity-seeking students of Reich's day. However, the
evangelical impulse to extend the environmental message
of global harmony stems from the same 1960s ideology.
The Northern White Empire is currently supporting the
enthusiastic but consensual domination of its global mode.

Ecotopia: *the vision of the 1970s*

No movement is complete without its utopian text. Liber-
tarians have Ayn Rand's *Atlas Shrugged*. William Morris's
News from Nowhere (tellingly enough cited on the cover of
Ecotopia) inspired early English socialists. Edward Abby's
The Monkey-Wrench Gang (discussed on p. 85) expressed a
tough and rumbustious reaction to the preservation conundrum. It was the individualist side of the ecotopian coin.
Ernest Callenbach's novel *Ecotopia* (1972) is the other. It is
a utopian tract closer to the virtuous socialist anarchists of
Ursula Le Guin than to the Heinlein-like wise old Doc and
armed veterans of Abby's work. A bestseller, still on recommended lists of ecological texts, Callenbach's is one of the
few serious and sustained attempts to show how an ecological utopia would work. Its emphases and omissions
typify the qualities which, I have argued, characterise the
American ecologist of this period: college educated, college
oriented, and imbued with a romantic collectivism owing
little to environmental problems.

Callenbach's *Ecotopia* is a picture of a seceded California,
where ecological principles reign. An ecologically sustain-

able economy has been introduced. Cars have been abolished, and long-range transport consists of electromagnetic trains, driverless and very fast, which are low-upkeep and very cheap to use. But this is no rural dream. Callenbach not only retains cities in his vision, but reclaims them from the motor car and the office block. His inner-city centres are densely inhabited, but the outer suburbs have been abandoned and razed. In their place rise up mini-cities, linked by rapid, free, clean train transport, but simultaneously committed to a certain level of self-sufficiency.

Some of the issues left unaddressed in the book, for example how the economy and political system would work, are significant to an understanding of the ecological vision, its ends and strategies. However, one thing is clear: in Ecotopia the ethos of the private individual, with his greed, his materialism, his competitiveness, has been displaced by a joyful and voluntary collectivity. This is important, because Callenbach is trying to show how the system of democratic participation and decentralisation, propounded as a vague target by left ecologists (that people will take their own decisions, come together democratically and decide issues on a local level), might work in practice. An analysis of *Ecotopia* helps to explain why so many of the demands of today's Green parties are not particularly environmental: the real target is a change of human power structures rather than saving the planet.

The vision is of a society where work is organised collectively, in brigades, and individuals live in communal buildings, according to their work and status. Journalists, for example, live in a kind of Writers' Union building. Washing-up and other tasks are shared. The hero, through whose eyes Ecotopia is seen, is a guest of the state. Career patterns in Ecotopia are left rather vague. Marriage patterns are the usual utopian ones: the extended family, children cared for by a group, women free to work, men and women living without ties or jealousy, extending sexual favours without hassle or recrimination but returning to their open relationships at will. There is a very low crime rate, and Callenbach hints that this is brought about partly through a kind of war game, rough but not fatal, which soaks up young male energies, as well as through hunting. The hunters, all male, travel by minibus and bicycle. The meat is divided up among the community and eaten. The hunters kill with bows and arrows, and are not squeamish about blood.

This at first sight implausible mixture of free bicycles and macho hunting rituals is countered by other values. For example, matriarchal arguments are represented in Ecotopia. The heroine runs a work brigade that manages a forest, but a place is found for some kind of tree worship or other occult tree-focused activity in her working day. She wears magic symbols round her neck, and is in touch with the currents of the earth, the spirit of the earth. She shows the hero the joys of making love in tune with earth's harmonies; she is brave and open in her emotions.

One of the key utopian beliefs is that human beings, once removed from the constraints of the market-place, the profit motive, competition and capitalism, have wonderful creative energies and powers. Ecotopia is an unusual utopia in that it does not show superman-type capacities and achievements unleashed by the new system. On the contrary, leisure and communal pursuits are prized. In Ecotopia work discipline is poor. When the hero takes part in simple tasks on the factory floor or in the kitchen, he is affectionately mocked for his orderly and rapid way of working. Ecotopians take time off as needed. They relax, chat, dance and sing a little, and are in general not subject to the economic and social constraints that induce haste in the world outside, or to the deadlines or the anxiety. This way of life is contrasted by Ecotopians with the stress and efficiency worship of the profit-oriented capitalist system that controls 'ordinary' American life. Callenbach wants to show here that it is possible to avoid worker alienation through introducing a self-organising, democratic work-brigade system, run on the same lines as a chores list in a shared house, without abandoning technology altogether. Ecotopia's political economy does not rest on handicrafts.

What a change of system does do is to unleash energy-efficient technologies. In forming Ecotopia, utilities are nationalised. They thereafter use resources more efficiently, while magical technologies are invented, or rediscovered, because malign corporations are no longer around to prevent their use. There is a cradle-to-grave health insurance scheme, which apparently is free at point of use. Doctors who were unsympathetic to the idea emigrated when health care was nationalised. Callenbach does not give a cost for this system, but it is able to provide a full-time nurse per patient, who not only cares for the patient's physical health but massages and nurtures him, part mother, part lover.

The taxation system is also left unexplained in the book, so it is not clear how this excellent and modern system would function without creating the problems of rationing health care or minimising medical advance found in nationalised health systems outside the USA. Callenbach's argument suggests that all human ills are caused by malevolent greed: allow the greedy doctors to emigrate, nationalise health care, and somehow it will be more extensive, better and affordable.

However, the key aspect of his utopia is the sustainable economy; no imports, no exports. The answer here is recycling. Human excreta and garbage are recycled. Sludge is produced from both and used to fertilise crops. Intensive farming is replaced by free-range farming, so that animal manure does not become a problem.

Food production is somewhat of a special case in this picture. The seceded state began by producing five times the food needed by the population. With food exports at an end, the issue is how to shrink food production, not how to extend it. The working week is cut to twenty hours, which is one reason why, according to Callenbach, the Ecotopians do not work very hard. Cutting working hours also helps to reduce food production. It is not shown how this affects farmers, but we learn that the 'surplus farm labor' is absorbed 'in construction work required by our recycling systems'. Other economies in food distribution are achieved through the resultant lack of choice, and by banning processed and packaged foods.

'Ban' is perhaps the wrong word here. In a key passage, the Assistant Manager of Food explains how it works. The lists of outlawed foods are inforced informally through moral persuasion. The lists are 'issued by study groups from consumer co-ops'. The committees 'operate with scientific advice, of the most sophisticated and independent type imaginable'. Scientists are not allowed to 'accept payments or favors from either state or private enterprise for any consultation or advice'. It is not explained how the scientists can afford to study products and give advice, but presumably they donate their time free.

Ecotopia is Spartan. There are no street signs and none of the paraphernalia of consumer society. No paint is used, because it is both wasteful and toxic, while clothes are expensive and made of recycled materials. Books are scarce, but are read eagerly and then lent to friends. It is perhaps typical of the ecologists' failure to address production problems that Callenbach does not explain the economics of publishing in such a society. Ecotopia permits delivery trucks (and garbage trucks), but there are no private cars. Shop supplies are moved in containers, which are loaded on to the electric trains and then on to electric trucks.

Non-American readers might understandably assume that to some degree *Ecotopia* glorifies experimental and laid-back Californian lifestyles, even though California's notoriously car-based standard of living is absent. Hunting, for example, is carried on using spring water and bicycle, not car and Coke. Ecotopians are physically very relaxed in a way associated with the cult of the body beautiful, sun worship and beach life. They are unembarrassed at making love in public, wear weird, hippie-type clothing, are inspired by the American Indians, and do yoga. But I suspect that a more important inspiration is at work, one that has been a constant theme in the work of the ecological gurus, and that is the campus. Callenbach's utopia expresses a remarkable faith in the virtues of nationalising industries, which seems to be entirely unaffected by any experience of government-controlled enterprises in practice outside the United States, or by any awareness of their history, but which is normal among the *Beamten* classes of all countries. A belief in collectivist virtue is common to the gurus of the time, and has characterised the Green movement increasingly since the 1970s, as the students and dons of that period continue their march through the institutions. The collective theme is indeed stressed strongly throughout the book, most tellingly when the hero learns that when Ecotopia became independent all waterfront properties were seized and made into 'water parks', while 'Beautiful exclusive estates' – a phrase wonderfully redolent of Jack London's lip-licking mixture of envy and hatred of the rich – 'were seized and turned into *fishing communes*, schools, hospitals, oceanographic and limnological institutes, museums of natural history' (Callenbach, 1977, p. 36; my italics).

The prevalence of university personnel among ecologists and environmentalists in the USA is striking to the outsider. Just as British socialism was deeply affected by the image of the paternalist Oxbridge college – meritocratic but just, selfless and unmaterialistic, geared to the higher values, with production only indirectly affected by reward – so American ecologism is a picture of a glorified campus. You work your way through it, but only part-time. Ecotopia's ideal factory is situated in what is unmistakably a campus environment:

Alviso has a cluttered collection of buildings, with trees everywhere. There are restaurants, a library, bakeries, a 'core store' selling groceries and clothes, small shops, even factories and workshops – all jumbled amid apartment buildings. These are generally of three or four stories, arranged around a central courtyard of the type that used to be common in Paris. . . . Though these structures are old-fashioned looking, they have pleasant small balconies, roof gardens, and verandas – often covered with plants, or even small trees. The apartments themselves are very large by our standards – with 10 or 15 rooms, to accommodate their communal living groups. (ibid., p. 24)

The lifestyle at Franklin's Cove, the 'press commune', with its communal kitchens and rap sessions, is the student ideal. The hero's emotional support system back home is criticised: 'You don't have a group of people to live with, to support you emotionally, to keep your collective life going on actively and strongly while you're apart . . . I find that very scary' (p. 32). The educational theme, so important to the ecological programme of persuasion without coercion, is stressed everywhere, as is the democratic one. The opening of a solar power plant is seen through the eyes of the hero as an informal occasion, where the workers who built the plant conduct the little ceremony of opening it: 'The woman describes the background of the plant – why it was needed here, how the people involved in the communities it would serve decided what kind of plant it should be, how some novel scientific developments got worked up' (p. 37).

The slightly Spartan nature of life in Ecotopia – and the Platonic comparison arises too through the lack of material privilege extended to the Guardians, or rulers – is mitigated by excellent home-grown food, cosy fires and stoves, and leisure. Clothes are expensive, bicycles free; other living costs are not described. The health service is good partly because California started with a highly educated population, and a high percentage of doctors and scientists, while 'medical schools were doubled in capacity after Independence' (p. 143).

Once the Protestant work ethic is abolished, production goes down, but it does not matter, because 'humans were meant to take their modest place in a seamless, stable-state web of living organisms, disturbing that web as little as possible' (p. 43), and the environmentally damaging impact of high-growth strategies is assumed in the book. Nonetheless, Ecotopians are seen as busy building mini-cities and railways, to replace road-linked suburban sprawl, and this seems to conflict with the aim of living in a 'seamless stable-state web of living organisms' without disturbing that web. Is the target of ecological enmity here disruption of the seamless ecological cycle, or is it something else? Callenbach has resource economics in mind, rather than biological preservation strategies. The aim of the stable state 'would mean sacrifice of present consumption, but it would ensure future survival – which became an almost religious objective, perhaps akin to earlier doctrines of "salvation"' (p. 44). (This is one of the few references to religion. With the decline of the Judaeo-Christian ethic, human life is taken more casually: terminally and critically ill patients are not kept alive: p. 143.)

Throughout the book the social and economic ideals of the left-ecologist are given priority over the ideal of preserving specific biological ecological webs. For example, in the financial chaos that follows independence, 'with the ensuing flight of capital' (and presumably the capitalists) 'most factories, farms and other productive facilities would fall into Ecotopian hands like ripe plums' (p. 44). So agriculture, that core of wasteful consumerism, is nationalised, retail chains are consolidated, and laws passed controlling oil industries and lumber companies.

The nationalisation of agriculture and expropriation of farms is hardly the most effective means of controlling big business. It is, however, a good way of controlling individualists, not to mention diminishing unnecessary food production. Socialists have tended to hang fire on it as a nostrum since the various experiences of agrarian collec-

tivisation in this century. Callenbach is perhaps exception-
ally dependent on the old left-anarchist dream of food
production without private property, but he follows ecological logic in one direction at least: that of opposition to
individualism and hostility to private property, which has
become a marked feature of 'Green' writing.

One omission should be mentioned. In his vision of the
steady-state economy, the work brigade, the leisured society where unnecessary consumption has been abolished
and everybody knows democratically just what consumption is necessary and what is not, the society where scientists and doctors work democratically with road-sweepers
and switch jobs with them, the spontaneous and creative
society where folk just rap together in the evening, the
society where would-be immigrants are sternly checked
and discouraged, in this steady-state economy – in *California*, of all states, there is no water shortage, no water
problem. Apart from describing the dynamiting of dams,
because they damage river ecology and prevent the people
enjoying boating facilities which are theirs by right,
Callenbach never mentions it. Yet California, and indeed
other states in the west of America, has such a major water
problem that it depends entirely on the goodwill of other
states for supplies, and is contemplating immigration re-
strictions because the water use capacity has already been
reached.

Callenbach's utopia represents the radical input into
American ecologism. There are of course ecologists who
support small farms, and who call for national self-sufficiency via independent peasants (Bramwell, 1989). Both
dreams are utopian, but people are entitled to their dreams.
However, the dream of collectivity, of local and democratic
participation, is now the one stressed by most Green political parties and theorists, especially those from or influenced
by the USA. One reason for this inconsistent preference for
an anti-individualist ethic may be, again, the influence of
student life, the matey peer-group world of the American

undergraduate, especially in the 1960s and 1970s. The ideal of student life, the easygoing commune, with sustenance and work organisation carried out by a paternal hand, its authority mitigated by democratic discussion, is that of the adolescent in secure accommodation. And non-left-ecologists have in their turn to answer the problem of how the ideal society is to be achieved. For the left-ecologist, the unfocused and unspecified dream of taking over farms and factories, making decisions about people's lives 'democratically', taking the reins of power into their own hands ('local communities regained control over all basic life systems': p. 62) is answer enough. At the back of the ecological mind, sometimes articulated, sometimes implied, is the belief that property was the original error and property rights sprang from territoriality. Settled agriculture necessitated control of territory. Callenbach's utopia exemplifies the belief that education and persuasion can and must undo the power structure of settled life, of societies developed from agriculture. *Property* can be undone. The restraints, the rigours, the brutalities of the agrarian life, the emphasis on the long-term view, the need for social control, can be abandoned. Abel can defeat Cain. Given free rein, nurturing earth will provide. A little bit of persuasion, even force, even strict control by a virtuous state, may be necessary to push humanity to that happy point; force can then be abandoned.

The hunter-gatherer vision that is so strong among ecologists (Hele-King, 1989; Oelschlaeger, 1991) is a utopian back-to-the-womb fantasy. This is seen very clearly at the close of Brian Aldiss's science fiction trilogy, *Helliconia*, where earth is invaded by bounding balloons, which provide homes, water, all life's requirements, for their human tenants, who can now live without disturbing earth's harmony.

I have tried in this section to trace similarities between the apparently very dissimilar writings of three of the gurus of the late 1960s and early 1970s. They have a love–hate relationship to the state, a dislike of elitist power structures, a desire for a more rationally organised economy, together with an authentic and unalienated relationship to the means of production, a preference for institutionalised living, and a desire for a warm, harmonious, yet spontaneous collective way of life. Aesthetics and those values which many environmentalists would assume drive Green thinking certainly take a back seat, although we do find a slightly self-indulgent 'why oh why' opposition to vulgarity, noise and loss of community.

This Platonic collectivism is based on an imaginary medieval ideal of community. Marx wrote in a similar vein, though he, and especially Engels, would also praise the industrial society for loosening the bonds of the old world. However, Marx was himself tapping a German criticism of modernity going back to Herder and the Counter-Enlightenment. German criticisms of the industrial and technological transformation of society continued during the nineteenth century, while German conservatives in the 1920s and 1930s battled with the appeal and the dangers of technology. Heidegger's 1930s belief that a 'New Man' was needed to live with modern technology was followed about fifteen years later by a cry of anguish at what the new technology had led to – the rape of the earth, the rise of consumerism. The new gurus took their values from Marcuse's works, *One-Dimensional Man*, and *Eros and Liberation*. Marcuse's romanticism about personal liberation and the eroticisation of work came straight from German radical conservatism of the 1920s and 1930s, as Habermas has pointed out (1985).

Since British opposition to urbanisation, technology and its concomitant alienation comes from a different source – we have our very own native proponents of this theme in the form of Ruskin and Morris – it is interesting that it took Schumacher and the imported American gurus to put Green thinking into orbit in the UK. Along with newly minted feminism and other 'soft' left causes, the indigenous flower was trampled on by the imported

American one, with its own peculiar flavour, traditions and contradictions.

The link between Ecotopia and radical thought is not surprising. The American radical tradition was active in the postmodernist criticism and pollution anxieties of the mid-1960s. The commune movement developed a Maoist puritanism which suited the anti-growth ethic of the time. The self-image of mainstream American environmentalism is that it is problem oriented rather than philosophical, and that it concentrates on trying to solve environmental problems, using the tried and tested American methods of local pressure, lobby groups and consumerism. Political ecology in the USA, however, if it ever finds party expression, seems likely to take shape in the direction outlined above, that of the radical utopian left.

The last frontier

The anti-capitalist input into ecologism is not necessarily linked with an especially revolutionary stance. On the contrary: as in Germany, the realists, the left-ideologues, want to work through the system, and support a 'Fabian' kind of reformism, despite the rhetoric of revolutionary opposition. The direct-action groups, such as Greenpeace, vital in attracting publicity and support, are among the most important organisations in American environmentalism, and include the eco-saboteurs, a group not yet very well known in Europe. The best-known 'ecoteur' group is Earth First!, which currently faces charges of conspiracy to commit sabotage acts in the United States. Its leader, Dave Foreman, was formerly president of the New Mexico chapter of the Young Libertarians and, according to one source, an aide to Barry Goldwater (Walker, 1990). Here, the liberal anarchist link referred to above seems to have flourished, a link exemplified in the book which inspired the philosophy of Earth First!, *The Monkey-Wrench Gang* (Abby, 1975).

This cult work, an enormous popular success in the USA, and dedicated to Ned Ludd (after whom the Luddites were named), describes a group of drinking, wenching, ageing tearaway heroes, including an inarticulate, foul-mouthed Vietnam veteran and a New Age bimbo along for the ride. Deciding to fight back against development and despoliation of the Western deserts, they begin by burning billboards, and graduate to blowing up dams. The group retires when the Vietnam vet is horribly killed, and the bimbo settles down with the ageing doctor and has children. The dead hero reappears at the end of the book in a slightly tongue-in-cheek romantic resurrection.

It is a sympathetic and funny portrayal, which helped to encourage direct action groups in the West of America. Edward Abby himself spent years working for the National Parks, and exploring the last wilderness areas left. He grew to detest the idea of the managed wilderness, with access roads for tourists, camping grounds and rangers. He believed that to experience the raw desert fully, it must be a challenge. He wanted nature to remain untouched for its own sake, but also as a kind of initiation ground, a rite of passage to instil courage and understanding in urban dwellers who would otherwise remain superficial and incomplete human beings. Not much nurturing mother earth here, but rather a celebration of the killer desert, the longed-for but feared adversary.

The ecoteur movement shows that the earlier split between Preservationists and Conservationists is still in being. The very size of federal intervention into land use has made the clash evident. Federal agencies own most of the North American wildernesses, including forests. They manage the forests: that is, they log and cut the timber and prevent fires. However, the ecological effects of clear-cutting are now held to be destructive. Similarly, the emphasis on fire prevention has been identified as too successful, as having prevented small, natural fires, which prevent over-growth of some species at the expense of others (seedling oaks become smothered by young pines, for example). When fires do occur they have a lot of undergrowth on which to feed. Debates continue to rage about how far the timber should be cut and how far dangerous animals should be controlled; all the problems indeed that arise when man mimics nature's stewardship.

Here science is of little help. Studies of the Yellowstone National Park and surrounding area have tended to show that the continuity theory of ecosystems is inadequate, that the history of a natural region over thousands of years is one of continuing change. What is at issue is a value judgement. Given the will, the species of the primordial forests could be preserved, whether or not the forests themselves were saved, but the argument for saving the ancient forests and wilderness areas is that the historical and aesthetic value of their pristine condition should come first.

Bio-regionalism: safety in stasis?

The desire for efficient resource use for environmental reasons has always been a major factor in Green thinking. It can be found in the proto-ecologists referred to in the Introduction, and is on the agenda of Green parties today. The belief that society can be reorganised in order to use resources more efficiently also has a long history among today's environmentalists. This is not as obvious a connection as it might seem. From thinking that things are not very well done to thinking that they must be better done is a step that requires a certain confidence in one's reforming abilities. The long list of works that argue for carefully calculated resource use assume the possibility of reform (Rifkin with Howard, 1980; Pearce et al., 1989). They assume that human society is sufficiently flexible and for-ward-looking to redraw its boundaries on new lines, owing more to natural features than to political, cultural or ethnic ones. They assume that the calculations necessary for this major step are reasonably correct. The pattern of fear, opti-mism and a mixture of enthusiasm and gung-ho vagueness is becoming a familiar one. It is exemplified in the idea of bio-regionalism, a recent American reworking of a familiar theme.

Bio-regionalism is based on two ideas. The first is that the earth is divided up into natural eco-regions, which can be broken down into smaller and smaller units, but which are all naturally self-contained geographical and biological units. If left alone by human hands, these areas will evolve to a steady-state position of maximum biological diversity consonant with the recycling of physical resources. This situation is known to plant scientists as 'climax', and the theory dates back to the biologist Frederic E. Clements (Worster, 1977).

The second idea is that human societies would be hap-pier, more self-sufficient, more diverse, the risk of conflict between peoples more contained, if they lived in a self-reliant way within these boundaries, and, if possible, within smaller units. The idea of self-reliance is a less rigor-ous one than the ideal of pure self-sufficiency. On the one hand bio-regionalism presumes that trade will be mini-mised – presumably to bare necessities, although the auth-or of a book on bio-regionalism, Kirkpatrick Sale, does not specify what these might be (Sale, 1985). On the other hand, the notion of a biological steady-state area assumes that a natural diversity of resources already exists.

These two ideas conflict. On the one hand, the bio-region would be drawn to take in a variety of climates, zones, and plant and animal life. Several writers who have become interested in bio-regionalism have drawn maps showing the ideal bio-region: it will produce a wide range of goods, and a variety of climatic zones:

Just as nature does not depend on trade, does not create elaborate networks of continental dependency, so the bio-region would find all its needed resources – for energy, food, shelter, clothing, craft, manufacture, luxury – within its own environment. . . . *It would be more stable, free from boom and bust cycles and distant political crises; it would be able to plan, to allocate its resources, to develop what it wants to develop at the safest pace*, in the most ecological manner. It would be more self-regarding, more cohesive, developing a sense of place, of community, of comradeship, and the pride that comes from stability, control, competence and independence. (Sale, 1984, p. 230; my italics)

An example of Sale's eco-region, the largest category, might be the 'Northeastern Hardwood'. He defines the bio-region as

part of the Earth's surface whose rough boundaries are determined by natural rather than by human dictates, distinguishable from other areas by attributes of flora, fauna, water, climate, soils and landforms, and the human settlements and cultures those attributes have given rise to. (Sale, 1984, pp. 226, 228)

On the other hand, if these areas are natural units, why do they have to be planned by human beings? Sale himself seems doubtful about trusting the inhabitants of the land. The dwellers need a little help from above. When it comes to determining boundaries he argues that 'the inhabitants, the dwellers in the land . . . will always know them best' (ibid., pp. 228–9). However, he also approvingly quotes Lewis Mumford, who wanted to resettle the entire population of the planet according to scientific principles (Bramwell, 1989, p. 81), and considers the Tennessee Valley Authority (TVA) as 'America's greatest – though in some respects most distorted –' experiment, although the TVA was hardly a product of the spontaneous understanding of dwellers in the land. Sale also approvingly cites Howard Odum's school of regionalism, begun in the 1930s. According to Odum, regionalism 'represents the philosophy and technique of self-help, self-development, and initiative in which each real unit is not only aided in, but is committed to the full development of its own resources and capacities' (Sale, 1984, p. 240).

The paradox that the bio-region is 'natural' and already exists, yet has to be reconstituted by human effort because it does not reflect existing political and settlement units, is a paradox that can be found in the works of many environmental reformers. They see human beings as interlopers in the natural system. One way of making them fit the ecosystem is to reorder their societies into regions self-sufficient in resources. But the sole reason for settling in self-sufficient areas is to consume that region's produce. Without human beings, there would be no need for bio-regions. Bio-regionalism is a way of minimising the disruptive capacities of humanity. But if some minimal trade within the bio-region is permitted, then why not outside the bio-region? The self-sufficient ideal assumes that the time, resources and energy used to move objects around from one bio-region to another are entirely wasted. There is no logical reasoning behind this. It is merely the anti-trade bias of the environmentalist. Indeed, there is a strong moral and political tinge to this concept. Sale does not discuss the carrying capacity of such regions. How many people could the self-sustaining bio-region, 'elemental and elegant' and stable in principle, support? We are not told. What would happen if population in one bio-region grew? We are not told. The purpose of the bio-region is to control human energies, and prevent change. It is the planner's dream, hiding under the guise of the spontaneous wishes of the dwellers in the land. It satisfies the desire to escape from the arbitrary booms and cycles of financial capitalism. A society based on bio-regional principles can plan, it is safe, stable, immune to faraway economic catastrophes. It is protected from the growth, the dynamic of capitalism. It is under control.

Sale argues that bio-regions to some extent already exist, and that America is already a federation of such regions. He also sees bio-regionalism as something to be encouraged from above. The desire for stability and stasis is a strong emotional given in Sale's writing, as it is in Schumacher's. Sale justifies it by his belief that nature is itself inherently stable, 'working towards . . . a balanced, harmonious, integrative state of maturity which, once reached, is maintained for prolonged periods' (ibid., p. 229). The unspoken corollary is that society should mimic nature's stability.

There were plant biologists in the 1930s, including A.G. Tansley, who disagreed with the climax theory of ecological development. Tansley argued that an ecology affected by human activities was just as valid an ecosystem as one undisturbed by human beings: he propounded what Worster calls 'a kind of environmental relativism' (Worster, 1977, p. 242; and see pp. 238–42). However, it is not necessary to take sides in this dispute to see that the Sale approach, although claiming to be the long-term, non-anthropocentric view, is in fact a very short-term viewpoint indeed, judged by standards other than that of the length of a human life. It is only in the short term that nature can be said to be stable, still less balanced, harmonious and so on. It is surprising how many people knowledgeable about earth's history fall into wishful thinking about benevolent nature. The developer of the 'climax' theory was affected by his experience of the North American grasslands, which seemed to form a self-renewing and stable ecosystem. But even he was thinking in terms of centuries, not of the millennia which ecologists claim to have as their time-frame. Not only have violent climatic changes swept the northern hemisphere in the last two million years, but oxygen-breathing life itself only exists as a result of a great catastrophe that wiped out previous forms of life. All species today have benefited from this 'crime'. We are all oxygen imperialists.

Apart from its role as model for a human ecosystem, bio-regionalism has a further importance. Nation-state boundaries are perceived by many environmentalists as wrong because they do not conform to eco-boundaries, and one consequence is that great environmentalist crime, trade. Nations are also a source of error because they do not offer a satisfying identity. They enforce homogeneity; they breed a central authority. Part of the attraction of bio-regionalism is that it offers an identity drawn from the earth around one, from one's ancestral roots in the land, an identity as deep and authentic as that of the American Indian, an inheritance unhappily denied to the white American.

Dissatisfaction with the nation state can lead to conflicting reactions: to support for globalism and for localism. Most ecologists support both. They support global means to enforce ecological principles, but they also support localism as the source of values, of diversity, of self-sufficiency. The idea of areas of North America seceding to form bio-regions has attracted many American thinkers who find the current system too centralised. Eco-libertarian Bob Nelson detects an urge to secede amongst many political forces in the United States, and sees this as a part of a pattern in Western history since the time of the Greeks. He calls this pattern 'Protestants' against 'Rome' (Nelson, 1991). Nelson's Protestants are not necessarily theological Protestants, but protesters; Rome is not necessarily the first and second state of that name, but represents any global international order. Protestants against Romans describes the revolt of the early Christians against the Roman Empire, as well as that of Protestants against Roman Catholics. Nelson's Protestants stir up trouble. They usually end up by overturning the previously stable order and becoming the new 'Rome' themselves.

Nelson sees the ecologists' demand for secession as a potentially destabilising demand, but one that could be made to work without complete revolution given sufficient thought and planning. Where Sale suggests a division of the USA based on 'natural' eco-regions, Nelson discusses the concept that the USA can be said to consist of at least nine 'nations' already, areas that in culture, trading patterns, geography and history have different pasts and presents (Garreau quoted in Nelson, 1991). He also points out that the centralised state has come under criticism from public choice economists like James Buchanan, whose moral critique of the centralised and rational state is surprisingly close to that of environmentalists (Buchanan, 1975, 1979) while global progressives, like Kenneth Boulding, see the state as hampering a move towards a world power (Boulding, 1970). Thus, the state is under attack from a wide range of thinkers, from libertarians to environmentalists. They all envisage decentralised seceded states under overarching global authorities.

Short of war or similar catastrophe, the USA is unlikely to break up. 'Strong' areas would find it hard to break away from declining areas. Romans do not voluntarily give up their empire. It may be that the arguments of environmentalists will do what other tensions and economic problems have failed to do: shatter the political homogeneity, the unified sovereignty of the United States of America.

CHAPTER FOUR

Germany 1970 to the Present

THE RECENT HISTORY of the Greens in Germany is interesting not only because of their importance as a delicate balancing weight in German politics, but because they are a fascinating and complex phenomenon. Through their size and access to finance, the German Greens have dominated European Green politics, and with their decline the Green cause has been severely weakened.

Environmentalism is just one of a range of programmes adopted by the German Greens. Some of these, such as restricting use of cars, follow naturally from a presumption of Greenness, but others, such as their approach to world politics, do not. Broad-spectrum Green issues have become submerged in a Green-flavoured package close to traditional left-liberal minority parties such as the FDP. Indeed, the one way up for the German Greens will be to replace the FDP as the safe minority partner in German coalition politics.

The German Green Party is the largest and best-financed Green party in Europe. Because of their development they are less Green than their image would suggest. Should this apparent divergence from the original ideology be considered a strength, a sign that the German Greens have become integrated into 'normal' politics, or as a weakness, a sign that institutionalised politics have defeated principles? Has this situation arisen because the Greens are now a developed political party, with the wings and the factions that entails, or is it something that goes back to the early days of their foundation?

From the point of view of party advantage, staking out the ground of the liberal, urban left has been a strength, giving the German Greens a full and 'normal' political programme, so that they can compete as a normal party on the political scene. However, it is certainly a weakness if looked at from the perspective of radical ecologists, and even from that of the concerned environmentalist.

Is there any special reason why German Greens should have adopted urban left-liberal ideas in so many areas? Policies of redistributive taxation and a social wage require a coercive and centralised authority to carry them out, and presume a society wealthy enough to afford this. Such policies, carried out amidst a scenario of declining affluence caused by forcible de-industrialisation, would require authoritarian rule to enforce them. Another anomalous German Green policy is that of welcoming refugees, especially from the Third World, which contradicts the deep Green belief that mobility of labour is bad, and that man must be rooted to the soil and to his immediate neighbourhood. It also contradicts plans for dissolving the nation state and creating bio-regions instead. Human mobility is almost as resource wasting, from the point of view of Green economics, as mobility of objects – and the merchandising of human labour is as distressing as the merchandising of things.

Some German Greens have claimed that they took up 'nice' or 'soft' left ideas to differentiate themselves firmly from old Fascist ideas: this anecdotal evidence is interesting but not conclusive, although it is true that, prior to the late 1970s, Greenness was seen as an incipiently sinister conservative or even Fascist idea in German thought (Bergmann, 1970; Pois, 1985). Looking at the genesis of the Green Party as outlined below, it seems to emerge as part of the alternative radical movement of the late 1970s. On the other hand, one would anticipate the shadow of the Fascist past to exert a powerful influence in Germany, more so than in other countries, and thereby possibly to mould ecological ideas into a different form. Perhaps the need for a deep ecological spiritual movement was so strong that it had to emerge, even though mutated by circumstances. If so, how far has the current programme mutated since the Green Party was founded?

The history of ecologism in Germany is dominated by the Green Party, but it received considerable stimulus from the American New Left, when their environmental interests reached Germany in the early 1970s. Furthermore, the Club of Rome report, *The Limits to Growth* (1972), was a major bestseller in Germany, while *Global Report 2000* (1982), the report on natural resources commissioned by US President Carter, is estimated to have sold up to 500,000 copies, and inspired the German government to commission a similar report. One of the first native environmental disaster books in Germany appeared in 1975, when Herbert Gruhl, a Christian Democrat representative, published *The Plundered Planet*, another bestseller. This work was even more pessimistic than Jünger's *Bussauer Manifesto*, but it hinted that, although irreversible damage had been done, there was still time to avert catastrophe. This somewhat dark attitude seems to have struck home to Germans of all ages and political orientations. Why? One writer thinks that the counter-culture was already in place, and looking for a cause. He sees a new 'alternative life-style' movement developing in the mid-1970s, with over 200 alternative newspapers appearing by 1980, and experimental communes developing, a lifestyle which drew on 'traditions established by some of the reform movements of the turn of the century' (Langhuth, 1986, p. 5). This is an interesting comparison, because one of the characteristics of reform movements in Germany, both at the turn of the century and during the 1920s, was their messianic tendency. Gurus appeared and an end-of-the-world mentality prevailed.

Possibly, then, the conjunction of counter-culture and ecologism was not inevitable. And if the counter-culture was already in place, looking for a cause, *The Limits to Growth* philosophy appeared to articulate the right one. In the words of one representative of the German left, Werner Huelsberg, 'the capitalist world order' was in crisis at the end of the 1960s. Then came the ecological critique:

> We had to smile, for there it was in black and white: capitalism was a crime against humanity . . . in order to achieve this growth it left corpses in its wake, destroyed our planet, blindly wasted our unrenewable resources, interfered with our basic needs (air, shelter, quiet and leisure) . . . and all the time it was leading us to catastrophe, a catastrophe which threatened to extinguish all higher life-forms on this planet. (Huelsberg, 1988, p. 9)

But this link of ecologism with Marxism came late (and in a somewhat 'species-ist' form) and was, in Huelsberg's words, 'stuck on to the old programme'. Indeed, the 1970s saw a battle between Marxists and pure ecologists. Ecologists were suspected of being unreliable in their predictions, inspired by a bourgeois protest against pollution, and 'eco-fascist because "the ecological protest movement, at least in Western Europe, almost always addressed its demands to the state"' (ibid.). Furthermore, the ecological movement was expected to deteriorate into a single-issue one; there appeared to be nothing on which to base a political party.

Nonetheless, the left tendency was there. Huelsberg quotes a survey of new social movements, which shows that participants in ecological movements in most West European countries (with the exception of Belgium) were more likely to identify themselves as left or centre-left. This table covered ecological, anti-nuclear and 'peace' causes, and although the methodology of these studies might raise an eyebrow or two (one wonders if those against 'peace' declared themselves for war?) the results tally with the surveys carried out early in the 1980s on post-materialist values (see Introduction). Huelsberg interprets this to mean that ecologism is post-materialist and left of centre, even though unrelated to the old working-class left politics. He also argues that the German Greens emerged directly from the anti-nuclear movement of the 1970s, with the Maoist revolutionaries he so ironically describes tagging along in the rear.

Citizens' initiatives

Citizens' Initiatives began as genuinely local groups, formed to protest against some local problem – the need for road crossings or action against threatened development, and the like. Their tactics of holding meetings, writing letters to newspapers and signing petitions seemed effective. By 1977 there were 300,000 members of Citizens' Initiatives formed expressly for environmental reasons (Langhuth, 1986, p. 7). By 1981, over 5 million Germans were organised in this way, more than the estimated 3 million British who were members of or affiliated with environmental pressure groups.

The first Citizens' Initiatives were middle class and apolitical in spirit, but the anti-nuclear movement (middle-class again but radical) rapidly came to dominate aims and procedures. The SPD (German Social Democratic Party) and the trade unions in Germany are, as is well known, integrated into the 'normal' political culture, and there were worker demonstrations in favour of nuclear power plants in 1976 and 1977 (Huelsberg, 1988, p. 61). The environmentalist middle class was politically marginalised by this worker preference for growth and jobs, which was similar to worker reaction in Italy to attempts by Italian environmentalists to close the chemical plant at Seveso, namely hostility and fear for their jobs.

Despite the reverence for nature lurking in the German psyche, environmental legislation was thin on the ground until the 1970s, while air and water pollution laws of 1974 and 1976–7 respectively were criticised as over-generous to the polluters and imprecise about permitted pollutant loads. In 1975 the Socialist–Liberal coalition dropped plans to cut factory emissions. It was not only the respect the German government held for wealth-creating German business that hampered legislation; the consensual nature of German society meant that even planning regulations were matters for discussion, arrangement and negotiation. The 1920s fashion for decentralisation had already led to polluting industries being sited in greenfield sites, and the results of this process were hard to reverse. But the German Greens were not at first interested in legislative reform.

> Protection of the environment today means more than eliminating or moderating some of the worst effects of the industrial system . . . environment, economic inequality, social injustice, and the growing dependence of the individual on the powers of the state . . . are essential features of the system. Our interest is not merely in the correction of errors and the elimination of unpleasant side-effects. Our goal is a more just, a freer and a more humane social order. (Huelsberg, 1988, p. 63)

So ran the 1980 manifesto of the anti-nuclear movement. Huelsberg explains this as the opening of a debate, 'which led, finally, to participation in elections and the formation of the Green Party' (ibid.). Unfortunately, the debate began well after opening hours. The New Left found, with some irritation, that others had got there before them. Conservatives, anthropologists, romantics, farmers – and worse. Luckily these interlopers could soon be seen off the scene. In Huelsberg's picturesque phrase, the ecological train belonged to the left but had been hijacked by the conservatives. The left had to run along the platform, jump on – and throw them off.

The birth of the Greens

In 1977 a local Green party, the Green List for the Protection of the Environment, appeared in Lower Saxony, gaining 3.9 per cent of the votes in the 1978 election (Langhuth, 1986, p. 8). Its chairman had been active in the Citizens' Initiative movement as an ecologist. 'Rainbow' lists, with a variety of radical groups (purple for feminist, black for anarchist and so on) soon appeared in other German cities. The example of Hamburg is particularly interesting, because there a split developed between the Rainbow List (allegedly subsidised by the German Communist Party) and the Green List for the Protection of the Environment. The Rainbow List received 3.5 per cent of the overall vote, and 18.2 per cent of 18- to 24-year-olds voted for it. The middle-class, more conservative environmental group received only 1 per cent of the vote. Clearly, there was something about the conservative variant of environmentalism that did not appeal to the voters. Contrariwise, ecologism could fit into a range of other left-political issues (Papadakis, 1984).

Perturbed by this, Herbert Gruhl, author of *The Plundered Planet* (1975) some five years before the Maoist left scampered into the field, formed Green Action for the Future. This group is still in existence, and issues its newsletters on greyish recycled paper, quite unlike the glossy Gucci greenness of the official Greens. Gruhl was actually a representative in the German Parliament at the time of his book, but had resigned from the Christian Democrat Union (CDU). He intended Green Action to be the nucleus of a Green party, but received little support from other wings of the Green movement.

The 1979 election played a role in spurring Greens on to action, as in Britain, although in this case it was the European parliamentary elections which were at issue. To stand for election, you had to be a member of a formally organised and nationally based party. Funds were available for parties that put up enough candidates. Although the organised anti-nuclear movement did not immediately support the idea of a Green party, there was a consensus among them that their traditional tactics were not effective. In March 1979 some 500 delegates met and agreed to fight in the European Parliament election as a Green Alliance. Delegates included members of Green Action, the Green List, the Action Association of Independent Germans, the Green List for the Protection of the Environment, the Green Action for the Future, and the Schleswig-Holstein Greens. An anthroposophical group called The Third Way was also present. Leading spirits of the new coalition included Petra Kelly, a feminist and pacifist EC civil servant; Herbert Gruhl, the former Christian Democrat; Joseph Beuys, the anarchist artist; and a representative of the Schleswig-Holstein Greens. The Hamburg anarchists and the Berlin Alternative List did not join in. Thus, contrary to the impression given by Huelsberg's account, the first Green delegates were not exclusively members of left groups.

The 3.2 per cent vote for the new alliance was disappointing, but it meant that the party now received state funding, and could set up offices and seek to attract members. The possibility of breaking the 5 per cent barrier and winning seats in Germany attracted back the left anarchists, and the alliance threatened to dissolve in fierce disagreement. Eventually they agreed on four basic concepts: 'ecology, social responsibility, grassroots democracy and non-violence' (Spretnak and Capra, 1985). These principles were sufficiently vague to avoid internal conflict, but proved too inexact for the formulation of policy. However, the non-violence clause would discourage extremists from right and left, the social responsibility clause appealed to more conservative supporters, and grassroots democracy could attract those who felt that the ruling consensus excluded them and their beliefs.

The growth of Green support in national German elections was slow, but seats in local (Land) elections came gradually. In 1983 the Greens won nearly thirty seats in the national election. Given the coalition-oriented nature of German politics, this meant the Greens faced hard choices about whether to soften their policies and co-operate with other parties or to try to 'go it alone'. The natural choice of partner was the German Social Democratic Party, the SPD, but here the German Greens ran into a problem that was to dog development of Green party politics elsewhere: conflict with 'old left' working-class supporters of left parties. Many of the SPD's supporters were trade unionists and worked in industry: hard-hat types who were not especially sympathetic to feminism, participatory democracy, pacifism and the dream of reversing industrialisation.

The openness of the Green Party's procedures meant that these conflicts were always visible. This may have held back support for them and prevented their becoming a mass party. On the other hand the ideal of constant debate and the right to disagree was so entrenched a principle that by going for unity over principle the Greens would have lost one of their major attractions – their image of sincerity and idealism.

The 1983 election brought the split between *Fundis* and *Realos* into play. Fundamentalists were against compromise, coalitions and reform. *Realos* were prepared for gradualism and were happy to use the political system to achieve a slow approximation to Green ends. They were hampered by the fact that the best-known, most dynamic and attractive Green figures were *Fundis*. One of them was Petra Kelly, who had left the SPD to join the Green Alliance in 1979, became chairman in 1980, and was speaker of the Green Parliamentary Group between March 1983 and March 1984. Kelly, a long-standing anti-nuclear campaigner, strengthened the Greens' image as a left-pacifist group. She opposed coalition with the SPD and continued her non-parliamentary activity in feminist and anti-nuclear movements. Like her many *Fundis* felt that it was more important to avoid softening their principles than to maintain a presence in Parliament (Parkin, 1989, p. 125).

The *Fundi* cause suffered a loss when Rudolf Bahro, a former East German and the best-known German Green writer and speaker, left the party and moved to the USA. Bahro opposed not only political compromise but the technological advances which meant that pollution could be cleaned up (Bahro, 1986; Dobson, 1990, p. 146). Environmentalism – the attempt to improve the environment and protect nature – was a betrayal of Green ideals, which to him meant a real reversal of energy use, growth and industrialisation. When a *Realo*, Joschka Fischer, became a minister in the Hesse Parliament in a Green–SPD coalition, the *Realos* appeared to have won the internal battle.

The Green Party continued to be riven by political differences. These were partly procedural – whether or not to support direct action, radical opposition – but also concerned the party's attitude to the increased 'Green awareness' of the German establishment, both business and political. Some Greens continued to support the conservative-oriented ecological programme of Gruhl's Green Action group, and did not see why the Greens should not form coalitions with parties other than the SPD. Support for the Greens was dented not so much by these ideological fissures, which after all reinforced their image as the party of thought and concern, but by the associated procedural wrangles. To avoid the cult of personality, the Greens introduced the rotation rule. This proved difficult to enforce, and has now been dropped, but the clash between ideal and reality in this instance was counterproductive: it worked against the Greens' public image. Apart from these internal contradictions, the dramatic political events of 1989–90 badly hit the Greens. For example, the Alternative List in

Berlin, one of the political success stories of the Greens, was effectively left-dominated, receiving many of its members from the Maoist left. Berlin was traditionally the home of the German left, and the Alternative List, recently elected to govern the *Land* Berlin in coalition with the SPD, found itself embarrassed by its attempts to appease the former East German government after reunification in 1990.

Reunification altered the balance of the disagreement over continued membership of NATO. While NATO was seen as an American-dominated Cold War artefact, many Greens, unsurprisingly in view of their roots in the German peace movement, assumed it was hostile to their values. The Green position was that capitalism and communism were outworn and discredited, and that the two opposed systems were gradually coming together. There were former anti-Cold Warriors who warned that Germany's commitment to NATO was not as bad as it seemed. While it held her back from *rapprochement* with the Soviet Union, it also tied her to a democratic West. These voices are now much louder. NATO is now seen as an alternative to a powerful and isolated Germany rampaging over the face of Central Europe. Thus, Green foreign policy, which relied on *détente* with the Soviet Union (not particularly through sympathy for its policies, but because of a dislike of confrontation and a belief in persuasion), was left incoherent.

The collapse of the communist regimes in Eastern Europe and the virtual bankruptcy of the Soviet economy made the Green ideal of converging systems harder to sustain. In particular, the failure of the 'soft left' East German dissident movement to gain popular support has been a blow to West German Greens. After all, the Czech Oto Sik in the 1970s and former East German Bahro in the 1980s talked of an alternative that would avoid the technocratic consequences and exploitative nature of capitalism and communism. And the environmental problems of Eastern Europe contributed to the downfall of communism there. Yet popular reaction in the new East European political systems has not been to vote for Green parties. While various reasons can be found for this, which will be discussed in a later section, the fact remains that when given the opportunity, the electorate in the East opted to vote for economic concerns or for nationalist causes, not for Green parties.

The result has been an embarrassment for the Greens. In West Berlin for example, the Berlin Academy of Sciences, formed as a counterweight to the East German Academy, came under attack from the Green Alternative List, who wanted it closed down because it might offend the East Germans. The Bill to do this was going through Parliament when the Berlin Wall came down. Now it is the East German Academy of Sciences which has been 'reconstituted' – together with the entire university structure.

Even before the collapse of communism in Eastern Europe, there were deep conflicts between the German Greens and East European ecological groups, who were active in dissent within their own countries. The Germans were more pacifist, opposed capitalist economics and the market, and saw the profit motive as encouraging greed and destruction. For them, spiritual criticism of Western society was paramount. Materialism was the evil. Not only did they hope that capitalism and communism would converge, they believed that a new third way was possible, a way which would lack the social coercion of one and the market coercion of the other. It is in this context that their left-anarchist origins emerged. The Polish and Czech ecologists did not link private property and the profit motive with environmental destruction. Neither did they attribute this destruction entirely to the political and economic domination of the Soviet Union. But in so far as human failings caused pollution, and not the evils of centralised planning and Soviet exploitation under the Comecon system, these were seen as sloth, cowardice, and above all a lack of individual responsibility.

The forced compromises between *Fundis* and *Realos*, urban left-wingers and nature-loving farmers, had another serious effect on the Greens. It made them much less Green. Their parliamentary activity during the 1980s was focused much more on drafting and proposing bills that they knew would never pass, and chasing up scandals in German financial and business life, than on environmental issues and nature protection. Their programmes tended to follow the clichés of the German New Left: the economy was always in crisis; capitalism was always on the verge of a crash. In the 1983 Green manifesto, out of forty-six pages, environmental and nature issues received five pages in total, the same space as that devoted to the Greens' policy of peace and partnership with the Third World peoples. The section on energy (under 'Taxes, Currency and Finance') consisted of rhetoric mixed with a half-digested survey of research programmes, surprising in a country with Germany's scholarly reputation and tradition, and ignored the body of serious work carried out on energy conservation.

According to the German Green argument, the industrialised nations, who represent only 30 per cent of the world's population, consume 85 per cent of the world's energy – although since the industrialised nations produce the industrialised goods, not to mention most of the world's food, that is hardly surprising. We learn that this demand for energy is pacified through 'energy imperialism', that is 'the exploitation of the energy reserves in foreign nations'. This explanation gives the impression that the West exploits poor countries and steals their energy, taking it at less than market price. Yet European countries (the USA is a net exporter of energy) import only two energy sources, coal and oil. The highest coal-exporting nations are Poland, Australia, South Africa and Venezuela, most of them hardly underdeveloped countries, still less countries that allow their coal to be sold at less than market price. Oil, too, comes from rich countries rather than poor ones. Far from the West exerting imperialism over the oil producers, the oil cartel forces the user nations to pay over the market price for it (that is what a cartel is for). Given the attempts by oil-producing nations to influence Western policies from time to time, energy imperialism seems more apt as a description of the behaviour of the oil producers than of the oil consumers. If the Greens were simply to say that they do not like the idea of moving energy resources from one nation for use in another, the emotional force of the statement would be much less and the irrationality of the fundamental objection displayed more clearly.

The manifesto's energy section argues that nuclear energy is a threat to democracy and human rights, that nuclear power plants are a military danger, and that there has been 'a constant increase in consumer consumption'. Whether or not nuclear power plants can easily be put to military uses is a matter for controversy: atomic energy specialists say it is impossible. Certainly, there is as yet no evidence that nuclear energy has threatened democracy and human rights. Countries with major nuclear energy programmes are as diverse as France, Czechoslovakia and Sweden. Since oil consumption has been reduced all over the developed world since 1973, and energy use in the developed world has flattened out, one can only marvel at the power of the political mind to believe rhetoric rather than reality. The Green answer to excessive energy use is a cumbersome mixture of nationalisation, banning of electricity advertising, energy taxes, banning of electric space and water heaters, banning of road freight, and a vague series of calls for alternative energy, which is described as being technically feasible and bound to replace fossil fuel use. Only the suggestion that legislation hindering alternative energy use should be abolished accords with the 'nice' and environmentally conscious image of the Greens.

But the German Green world-view was always directed at something other than environmental or ecological issues. For example, it always had a claim to a smaller, gentler and more consensual lifestyle. It has always criticised hierarchical culture, culture which formalises the distinction between us and them, and which the Greens

perceive as embodied and stratified in museums and elaborate and expensive theatres.

A real search for authenticity would take this criticism to
its logical conclusion: support spontaneity, destroy the
icons. Greens propose instead subsidies for travelling theatres and mobile exhibitions, as well as the directing of
culture away from city centres to suburban and rural areas.
Cultural activities and subsidies should be directed especially towards women, the elderly, children, ex-prisoners
and former drug addicts – an unfortunate grouping of
categories. One of the Green manifestos is illustrated by a
photograph of a vision of Green culture, which the Greens
would like to substitute for the hierarchical formality of
Western culture.

Let us take this iconography seriously. In the photo two
banjo-playing men stand in a street, singing and smiling at
each other. Several embarrassed youths and women and
one old-age pensioner stand around them. This is an example of the warm, folksy, local culture we are called on to
encourage. In a country with the most vigorous local
musicmaking tradition in the world, a country where the
smallest town has its own wonderful musicmaking talents,
this photograph and its cheerful, consensual banality is a
monument to something very strange. To some extent it
indicates a hostility to effort, to meaning and to historical
experience, in the guise of community spirit.

If 'authenticity' was intended, the destruction of the
structures that perpetuate dead formalities, that stifle the
new, all anarchists would rejoice. The year 1848 saw Dres-
den burn, an experience from which Wagner escaped to
dream of Greek theatres open to all the people, with an art
form equally accessible to all. But here the tradition of left-
anarchist ideology is hidden behind a terrible benignity, not
burnings. Sunflowers and banjos; this simplification of
Green ideology towards consensus and 'niceness' was accompanied in the late 1980s by the rise of the old German
problems; with unification and the fall of the German

Greens, the vacuity of the images was laid bare. Greens who had talked of searching for the German soul now produced logos of a childlike character, ideographs that perhaps reflected the desire to escape from the soul they saw.

At the time of writing, the new vacuum of power to the East, and the vast costs of reunification, have produced new options as well as new problems in economics and foreign policy. These have to be tackled through trade and through the development of something unneeded since the war, a foreign policy. German identity will be redefined by conventional means over the 1990s: through new trade empires, through the rethinking of taxation and welfare systems and of corporate models, with a view to optimising national survival. All the post-colonial issues of the 1920s have returned: the potential trade links with the Baltic States, the apparently unstoppable urge to pour money into Russia in an attempt to buy stability and safety, with Kaliningrad instead of the Polish corridor the next bone of contention. The European Union is not the League of Nations, and Germany is unlikely to leave it, but she no longer needs it for legitimacy. The European Union's potential in foreign policy is clearly limited, and Germany is unlikely to remain milch cow and subservient partner in a French-dominated system.

With the stirring of power, the era of sunflowers and banjos is over.

It is in this context that the saddest footnote to the era should be considered: the deaths of both Petra Kelly and Gert Bastian.

Requiem

It was like a play by David Hare rewritten by Pat Buchanan. Three weeks the corpses of two radical activists lay undis-

covered in a flat, one, at least, murdered. No visitors, no
calls. Like old ladies who had had a fall in the bath, they lay
there. It is this fact that sticks in the throat most of all.
Truly, Petra Kelly and Gert Bastian had become alienated
from the world. In C.S. Sisson's sonnet which begins, 'How
can it be that you are gone from me, Everyone in the
world?', he writes:

> In the end those to whom one cannot speak
> Cannot be heard, and that is my condition.

The estrangement of this couple from the world implies
that eventually, for Bastian, Kelly was no longer to be
heard.

But the political is the personal. Fellow Greens found it
hard to believe that plain murder and suicide could have
wiped out this committed, passionate and dynamic couple.
A group of Green sympathisers, including Kirkpatrick Sale,
the American writer on bio-regionalism, called for further
police inquiries, because they could not believe in the obvious interpretation: that a depressed and ageing man, warrior and peacenik, took the life of his devoted lover and
companion. Whatever the truth of this horrible event, the
semi-gloating tone of media coverage of it combined a
displeasing salaciousness with a degree of *Schadenfreude*.
Although Petra Kelly had not aroused any particular hostility when alive, in death, it seemed, she became an alienated figure, as if being a murder victim demonstrated
retrospective alienation. If you buck the system, it was
implied, this is how you end up. With someone who threw
her personality so wholeheartedly into her politics, it is
perhaps understandable that interest should focus on
the surprising link with Bastian, former general, former
Second World War warrior. Kelly was partnered with Sicco
Mansholt for some years, too, so she obviously was drawn
to tough but caring father figures. But so what?

Kelly's politics were not in fact shockingly radical, being

mainstream American liberal; they were a culture shock only in 1970s Germany, new to the blend of women's rights, egalitarianism, convergence with the USSR and anti-nuclear activism that characterised 1960s America. Kelly brought charismatic politics over from her American experience, her dehistoricising completed by a period at the bright new European Commission. The German Greens formed themselves in her mould, more than in that of any other activist, and when she refused the rotation rule that the Greens brought in she was close to a final break with them. By then, ironically, her focus on human rights was remote from the anti-anthropocentric nature of fundamental Green ideologues.

Friends of Kelly think that she was stricken by the take-over by the mainstream parties of so many Green ideas and principles. But that has happened in most countries where there was a strong Green vote; what could be absorbed into 'normal' political life was taken and the Greens were left with the residue – radical, romantic and wrong. Kelly may well have been disillusioned by the form that this absorption took in Germany, where high environmental product standards were used as a weapon in an undeclared trade war, where sanctimonious postures about pollution were combined with the near-illegal selling to Eastern Europe of pesticides and toxic waste that were illegal in Germany. But for an old campaigner like Kelly this would surely have been all part of the predictable process. A more serious blow was unification and its results. The German Green movement had collected together all the anguish about the German soul, the German mission, that came from the violence of Germany's division. 'Over there' – *darüber* – were the dissident socialists, the anti-materialists like Bahro. Convergence theory was fundamental to the German Greens: the belief that capitalism and communism were both bound to converge. The reality was far otherwise. The collapse of the European communist bloc revealed the ideological void 'over there', and also the threat of a virulent reaction against Western liberalism, while events since then have revealed the inadequacy of the Western ethic. Eastern Greens wanted more individualism, not less, more material benefits, not fewer, since they attributed their pollution problems to that very state control and collectivist idealism that the Western Greens seemed to support.

Theories about the double death are up for grabs. But Sisson's sonnet ends with these words:

> if I do not reach the
> Outer shell of the world, still I may
> Enter into the substance of a leaf.

That is as good a requiem for Petra Kelly as any other.

CHAPTER FIVE

Britain 1970 to the Present

The background to the Green Party

THE POST-WAR EVOLUTION of Green philosophy in Britain was notable for its non-party political nature. The scientists and activists who campaigned on ecological issues arrived at their beliefs from a variety of disciplines, and experienced considerable cross-fertilisation of their ideas. After 1945 the main issues stimulating ecological concern were resource fears; population fears; anxiety over destruction of the rural landscape, trees, plants and animal species; air pollution; water pollution, especially from agriculture; and nuclear power. Of these issues, only air pollution received a legislative response in the form of the Clean Air Act, until water privatisation and EC pressure forced the British government into taking water pollution seriously. Unlike the USA with its 1960s legislation, no catch-all environmental Act was passed until 1990.

This delayed response requires some discussion and explanation. In part, it was due to the low priority accorded environmental issues, while this was to some extent due to the unresponsive management and run-down nature of the British nationalised sector. Water protection, for example, limped along, hampered by a demoralised inspectorate and a system of regulation that tried to avoid criminalising the issue, and concentrated on prevention rather than punishment. In the case of water pollution caused by agricultural activities – slurry run-off and nitrate pollution – it is difficult to identify the agents as individuals, and British policy encouraged intensive farming and the use of chemical fertilisers and pesticides. Further, environmentalists' claims were in general weakened by the public perception of them as cranky, right wing, middle class, irrelevant and 'selfish'. If you stood in the way of progress, you were selfish. The term NIMBY, meaning 'not in my back yard', had not yet been invented, but if it had it would have been used continuously. Local conservationism was seen as unsustainable. It took the globalisation of environmental issues to make them acceptable, to make them seem relevant.

Two wings of environmentalism have been observed in other countries: the conservative, rural preservation, low-membership, low-profile groups, and the mass-membership left-oriented groups with political ambitions. In Britain the latter category did not appear until well into the 1970s. The environmental constituency was, however, itself divided. One element believed in the desirability of organic and rural values, focused on organic agriculture and the need to go back to the land. This movement was hampered by its conservative image, and was attacked by establishment groupings in agribusiness, medicine and the chemicals industry. Their peripheral links with other 'alternative' ideas, such as occultism or anthroposophy, were stressed by their enemies, who were dominated by the scientific, rationally minded constituencies.

But there was also an economic ecologism, based on the undesirability of the growth ethic, and supported by economists such as David Pearce, who was to become prominent as an environmental economist in the late 1980s and to complain that he and others were 'saying all this in the early 1970s' (Pearce *et al.*, 1989). Their arguments were based on a critique of what was later to be called the 'non-sustainable' society, and stressed the logical impossibility of indefinitely sustained high growth rates. Given exploding populations in the developing world, and their inevitable

demand for an increased standard of living, the concomitant resource use would be exponential, not arithmetical.
To some extent qualitative factors were added into the
equation. Overpopulation plus a demand for rising standards of living also meant a decline in the quality of life,
through overcrowding, noise and pollution. One answer
was the idea of inter-generational resource allocation. The
theory was that finite resources, such as coal and oil, should
be shared with future generations. So, when calculating the
costs of energy use, some element should be introduced to
indicate the cost to the future of being deprived of this
valuable source of energy.

Population and finite resource fears were hampered by
the sensitivity of the population argument. Apart from
Hong Kong and a few other small islands, the most densely
populated areas per hectare were in Europe, where population growth was stable. The population explosion was in
the Third World, which the international establishment
was committed to developing and helping to bring up to the
same standard of living as the First World, while refraining
from political interference. To urge population restraint
within an ideological framework of generous non-inter-
ference was tricky and ineffective, and could hardly be
made into an idea with mass appeal without evoking
memories of the 'Yellow Peril' and other pre-1914 fears of
domination by non-white populations.

Perhaps another reason for the failure of this idea to
attract mass support is that for women one generation
away from the bondage of wash tubs, heating water by
hand on coal fires and hand-sweeping, attacks (usually by
male authors) on modern labour-saving machinery, wash-
ing machines and Hoovers were not appealing. If these
issues could be brought into a framework of global com-
passion they would be much more acceptable. When added
to anxieties about nuclear power, they would provide a convincing framework of anti-industrial, anti-Western beliefs.

This important step towards seeing environmental problems in a worldwide context, and calling on the inter-

national agencies to do something, was stimulated by *Only One Earth*, a report to the United Nations produced in book form by Barbara Ward and René Dubos. Coming just after the appearance of *The Limits to Growth* (Meadows *et al.*, 1972) this work won over many in the rational scientist camp who had been unaffected by the scientistic mysticism of Teilhard Chardin and his followers, and who attacked futurology and mysticism alike. The element of 'holism' in Ward's vision seemed more acceptable than Chardin's because it appeared in a framework of discussion about resources. The right cue-words brought about a rapid conversion among the scientific constituency since, after all, scientists need values and have non-rational parts to their psyches like the rest of us, and their critical mechanism can be unhinged. The history of political ecology has thrown up, and indeed depends on, many scientists who, outside their own disciplines, appeared to lack any saving element of reductionism, any critical analysis (Bramwell, 1989). The process can be seen clearly in the inclusion of Nobel Prize-winner Peter Medawar in the apparently wholesale conversion of a scientific elite to Doomsday thinking. In his book of essays, *Pluto's Republic* (1982), Medawar looked at Barbara Ward's book, and then at an attempt to counter her prophecies. He was struck by the book's appreciation of scientific development and understanding:

> We are left in no doubt that it is towards science and an enlightened technology founded upon it that we must turn to find a remedy for our present condition. But that will hardly be enough unless we contrive to develop for the earth as a whole the deep and passionate sense of allegiance which youngsters are brought up to feel for their birthplace, school or nation. (Medawar, 1982, p. 279)

Medawar was convinced by the book's message, that

> The two worlds of man – the biosphere of his inheritance, the technosphere of his creation – are out of balance, indeed

potentially in deep conflict. And man is in the middle. This is the hinge of history at which we stand, the door of the future opening onto a crisis more sudden, more global, more inescapable and more bewildering than any ever encountered by the human species and one which will take decisive shape within the life-span of children who are already born. (quoted in Medawar, 1982, p. 280)

Interestingly, the actual message is vague although frightening. Man stands between two unbalanced and potentially opposed worlds, one of his own making, one he was born with; and at a time when we face a crisis within some eighty years which will be sudden, global, inescapable and bewildering, more so than any humanity has faced before.

Medawar believed this, and his reasons for doing so are significant. In a review of John Maddox's *The Doomsday Syndrome* (1972), a book which attacked the Doomsday prophecies of the time, Medawar described Maddox's critique of the anti-growth position as 'mischievous'. He stated his support for the 'principal contention' of works like Goldsmith's *Blueprint for Survival*.

A small fraction of the world can continue to improve its standard of living, but only at the expense of degrading or failing to improve the lot of the remainder. In spite of their lesser numbers, the highly industrialised fraction of the world consumes 80 per cent of the world's energy and raw materials. Although the *populations of both worlds are increasing at a dangerous rate*, the growth of the industrialised moiety represents by far the greater danger . . . resources are finite. . . . Worse still, the cycles of nature are being disrupted so that natural regenerative processes are impeded or stopped altogether. (Ibid., p. 281)

This summary is the work of a man who wants to believe that the industrialised West is guilty, is exploiting the Third

World and unfairly expropriating the earth's resources. Yet, with some of the energy they use, Europe and the USA produce a food surplus. They are major food exporters, thus helping to feed the less industrialised world, besides until the last two decades producing much more of their manufactured goods. The belief that the industrialised world's population was increasing at a dangerous rate could have been checked with any yearbook, and found to be nonsense.

Medawar's reaction to Maddox's attacks on the factual basis of Doomsday thinking was to point out that *Blueprint for Survival* had had such an impact

> partly because its thesis was endorsed (with unspecified reservations) by upwards of a dozen assorted scientists, economists or public figures. Although it could be (and was) objected that very few of them had any expert knowledge of the matters under discussion, it could be said in defence that the signatories were united by a fearful anxiety . . . about the future of the planet. (Ibid.)

The fearful anxiety, even though it be of 'a dozen assorted scientists, economists and public figures' utterly ignorant of the matters under discussion, was enough to convince Medawar of the *factual* basis of their fears. This article was written in 1972; by 1981 he was still sufficiently convinced to leave the piece in his collected essays, despite the fact that the ten to fifteen-year-old prophecies of famine, immiseration, mass starvation, disease and congestion caused by inequitable resource use and population growth had not yet come true. John McCormick has calculated that out of five major prophecies of ecological doom made in 1969, only one related to environmental issues – the decline of the Peruvian anchovy fishing industry – came true. Where 50 million people were predicted to die each year from famine, current estimates are that over twenty years some 12 million had died from famine. Important marine life had not become extinct by 1979, and so on (McCormick, 1989, pp. 79–80).

Despite the effort and resources put into promulgating the doctrine of the Doomsday syndrome, the effect was mainly on international politics, via the conversion of a significant part of the intelligentsia. The finite resource argument was too obscure to appeal to most people, and many could see the logical snag: if you were rationing resources for an unlimited number of future generations (and nobody knew how many there were going to be, how far into the future the precious oil would have to stretch), then the rate of use had to drop to virtually zero. Finite resources were finite, no matter what you did with them, or how you stretched them. This conclusion was politically unacceptable at the time, and was not stated by ecologists. The Doomsday scenarios continued. *The Limits to Growth* (Meadows *et al.*, 1972) argued that by the year 2000 food shortages, environmental pollution and finite resource exhaustion would produce mass catastrophe. The physical limits to growth had already been reached, and indeed overreached. As contemporary critics pointed out, they assumed an extrapolation from present trends, taking the worst possible estimate of such trends, and assumed no technological or social change, no new resource discoveries. In 1982 *Global Report 2000* pointed out that none of the team who compiled the *Limits to Growth* report was a social scientist. Nor were they economic historians or experts in the dynamics of the developing countries (McCormick, 1989, p. 81). Despite such methodological doubts, these prophecies were accepted as a charter for global planning by global agencies.

The oil crisis of 1973–4 was a shock to oil users, and a lesson on the dangers of relying on cheap energy. To the environmental lobby it was manna indeed. Yet reactions to this sudden and catastrophic price rise showed the technological and social resilience of advanced industrial societies.

Although finite resources, such as oil, could be rationed by price by their owners to make the resource last, the lesson to be learnt from the experience remained uncertain. Environmentalists became fond of using the parable of the last match in the matchbox. According to this, you used up your matches until you had one left, then found that disaster was upon you. But in real life, you would go out and buy a new box when you had only five left, and if it appeared that no matches would ever again be made, you would look around for an alternative. What is wrong with this procedure? To stop using matches and bewail the complexity of modern technology would not be a helpful alternative. Oil (to escape from the parallel) *is* finite, and no matter how much is discovered (and currently more is being discovered each year than is being consumed) it will, one day, other things being equal, be used up. Yet to stop using oil because of this fact would not help matters. Rationing finite resources for the benefit of future generations is a frequent answer. However, economists have had to create entirely arbitrary methodologies for putting this idea into action, such as the suggestion that three future generations should be considered when 'rationing' the oil. Delaying the end of oil for three generations would only prolong the agony, whereas the oil price rise of the 1970s did force substantial cuts in energy use all over the developed world. The extent of energy conservation is surprising: energy efficiency improved in all Western countries between 1973–4 and 1980, the use/GNP ratio declining from 103 to 61 for Japan, and to below 90 for the US. Countries which had introduced measures to ration oil use saw a lesser decline in the use/GNP ratio (Singer, 1984, p. 344).

Nonetheless, the oil crisis of 1973–4 seemed to come as a vindication of the Doomsday scenario. In imitation of finitude, but through the old-fashioned mechanisms of the cartel, the oil price went up four times, and the energy crisis burst upon the world. Many people retain a vague feeling that it was this event that precipitated widespread ecological concern, although such concern pre-dated the oil crisis by some years. The explosive mixture of conservative values, enshrined in an agriculturally based world-view, and the more radical finite resource economists and scientists, had already formed, and it was this mixture that was to dominate British environmentalism for the next decade and a half.

The Green Party

In 1972 Edward Goldsmith produced his *Blueprint for Survival*. It summarised the ecologists' arguments, and went further: it recommended action, to be taken by governments and by individuals. There was a timetable. First on the list was the establishment of a national population service. This was followed by taxation of resources; population targets, to be enforced by law; and recycling grants. Six points of the manifesto dealt with food and organic farming, and one with roads, arguing that road-building should cease. The emphasis on agriculture suggests the influence of the older more rurally oriented ecological groups. The historian of the European Green parties, Sara Parkin, writes that with the publication of *Blueprint* Goldsmith intended to start a movement for survival, 'which would be a coalition of environmental groups, starting with those who had already expressed support for the *Blueprint*; Friends of the Earth, the Henry Doubleday Research Association, the Soil Association and Survival International' (Parkin, 1989, p. 214).

The Movement for Survival was launched in 1972, 'committed "to act at a national level and if need be assume political status and contest the next general election"' (Lowe and Goyder, 1983, p. 72). However, it was hampered from the start by disagreements about strategy. Some contributors to *Blueprint* thought it would be a mistake to develop a political party, and preferred a lobbying group, which would try to influence MPs. Indeed, existing lobbying groups did fear that their influence would be damaged by the existence of a formal, and militant political party. However, a group among the original supporters thought that party activity was preferable. They argued that lobbyists would always be in a weak position *vis-à-vis* the established parties, always liable to be 'bought off' and thus defused (Parkin, 1989, p. 215). Meanwhile, the founders of the Movement for Survival were more concerned about imminent collapse than the slow road to political power.

In the end, the Ecology Party (originally called 'People') was formed in 1973 by a small group of activists based in Coventry. This might have seemed a promising time to start an ecology party. The ruling parties were both in trouble. Edward Heath's Conservative government, which had come to office in 1970 with what, with hindsight, might be described as Thatcherite promises of reducing inflation, controlling unions and rolling back the state, had performed its famous 'U-turn', which encouraged union militancy without pacifying Conservative supporters. The two Labour governments of the 1960s had left many disillusioned. The time was right for a new third party, a party that escaped the old class-based framework and called for a new way.

Yet the Ecology Party failed to make headway. At first sight, the reason seems obvious: Britain's two-party, first-past-the-post system would never allow success to a third party. The 15 per cent vote for the British Green Party in the European election of 1987 was widely taken as a gesture of support for a party that could not gain seats at home, in an election where Britain might eventually allow proportional representation in some form. However, the Ecology Party did begin to gain seats at local level but not at national level, despite the fact that local government elections also operate a first-past-the-post system. This suggests that national politics were perceived by voters as an unsuitable vehicle for environmental policies.

Some writers have argued that the Ecology Party was hampered by its relation to the left (Lowe and Ruedig, 1986b, pp. 271–2), and the left image of British Greens has struck other commentators (Dani, 1989). Not only did this prevent voters taking them seriously as committed environmentalists, but organisation and strategy formation was held back by an ideological split. Tension appeared between the more conservative Goldsmith supporters, who wanted holistic science, a return to strong family values and who were opposed to feminism and open national boundaries, and the left groups, such as the Socialist Environmental Resources Association (Lowe and Ruedig, 1986b, p. 272), that began to take up environmental issues during the early 1970s, and were committed to the 'soft left' policies of pacifism, feminism and a basic wage.

This may account for the Ecology Party's slow progress during the 1970s. However, with the emergence of new leaders like Jonathan Porritt in the late 1970s the party seemed to have found a more appealing image. They decided to field candidates at the 1979 general election, thereby gaining media time and attention, but they obtained only a very small proportion of the vote. In 1983 the proportion of votes cast was up to 1 per cent of the voter population. Low as this was, the German Greens had had only 1.5 per cent in 1980 (Lowe and Ruedig, 1986b, p. 266). By 1984 membership had doubled, to 5,000. By 1985, when the Ecology Party changed its name to the Green Party, partly as a result of the success of the German Green Party, the term 'Green' was becoming widely known. Yet still the vital breakthrough did not occur.

One reason was the resounding success of the integration of environmentalism into Britain's lobbying system. An estimated 4 million people belonged to environmental pressure groups of one kind or another by 1986, and that figure does not allow for a relative weighting for the influential economists and journalists who supported environmental ideas. Apart from the fact that potential party members may have decided that pressure group action was more relevant, more fun and more likely to succeed, a large proportion (80 per cent in 1984) of Green Party members themselves belonged to some environmental organisation, while nearly half were members of the Campaign for Nuclear Disarmament (Parkin, 1989, p. 216).

Yet with all these caveats, it may be that the real reason why Green parties in Britain and elsewhere failed to become potential governing parties lies elsewhere. Both the ideological structure and the accidental dependence on the example of German Greens were to prove disasters for the national Green parties when, towards the end of the 1980s, the environmental movement abruptly moved out of the shadows into the forefront of political consciousness throughout the West.

PART TWO

Strategies

CHAPTER SIX

Politics and Tactics

ECOLOGICAL STRATEGIES CAN be divided into three groups: civil disobedience, the pursuit of political power on a national level, and the pursuit of influence on an international level. Of these three, civil disobedience is the strategy most consonant with radical ecologism, although it can be confrontational or non-confrontational: inspired by Gandhi or by Japanese martial arts, violent or non-violent, interventionist or focused on the inner spiritual life. I will argue later, though, that even the apparently innocuous meditative form of deep ecology is a contradiction, because it is oriented towards empowering the meditator.

Here I want to look at the varying experiences of ecological pressure groups in different countries, to try to understand why direct action should be favoured more in some countries than in others. I will examine Italian environmentalism as a special case.

The doctrine of civil disobedience has been most widely and successfully cultivated by Scandinavian ecologists. They find it an effective way of putting pressure on the state, perhaps because Scandinavian governments are prepared to play by the same rules of consensus and good will. The country where ecological activism is most violent is America, followed by Italy, Germany and, in the form of animal rights activism, Britain. It may be significant that the American group, the 'ecoteurs', stems not from a left-oriented tradition but from an elitist anarchist tradition that lies behind a certain kind of American liberalism (the Tom

Paine variety). This tradition fits in with the gung-ho backwoodsman argument used by Aldo Leopold to justify retaining wilderness areas – that the special characteristics of American democracy rest on the socialising impact of the wilderness. Why should eco-terrorism have got off the ground in the USA, and not in European countries? The answer may lie not so much in the varying experience of the *left*, much as direct action on this issue is associated with the far left, as in the fact that the function of political conscience is associated with individualism in the USA, but with communal activism in We tern Europe.

In Germany, the protest movement in the late 1960s was not a spontaneous cry of alienation but a highly organised phenomenon. Here is a telling description by a former sympathiser, now a left-Green, whom I quoted earlier.

> The chairman (or was he called general secretary) made his way to the platform – blond with a touch of grey, tight lips, the prototype of the ascetic, professional revolutionary. Without a hint of emotion he would read from the prepared text, without excitement, without enthusiasm, without hatred, without a change of tone. Straight from the head. As the code words fell, the hall would spring to its feet, clenched fists in the air. 'Up with' . . . 'Down with' . . . 'Long live' . . . always exactly three times, as if the ghost of Stalin hovered over the crowd, directing them with his invisible baton . . . zombie-like obedience. (Huelsberg, 1988, p. 8)

These tidy radicals then moved to 'the long march through the institutions', to use the phrase made famous by Rudi Dutschke (the famous radical who later became a Lutheran priest and ardent Green supporter). The constituency of the German Greens is firmly composed of those who did successfully march through the institutions, so that left-wing radicalism has been defused by the fact that a left-wing and not very ecologically oriented party has won a voice in local and national parliaments and the media, and has created channels for winning and exercising power.

In Germany former hard-line Marxists and Red Army supporters are now *Realos*, or realists, and oppose the *Fundis*, or anti-reformers. In Germany there is no need for eco-terrorism, or civil-rights-type activities. The Citizens' Initiatives movement of the 1970s, which could have joined forces with single-issue anti-nuclear groups, found a political voice in the Green Party.

Italy presents a different problem, and needs to be considered separately.

The strange case of the Italian Greens

Italy is the clearest example yet of the twofold division of ecologism into conservative and radical, but the two wings ended by merging into a more enthusiastic co-operation than anywhere else! An early interest in ecology in Italy can be glimpsed in organisations of biologists and scientists in the early decades of the century. The Fascist Party, which ruled Italy from the early 1920s to 1943, had a pro-peasant wing, which tried to promote peasant values and collected peasant songs and folk tales, but which lacked the other defining characteristics of the ecologist discussed earlier in this book.

Environmental concern surfaced in the late 1950s in a conservative form with groups like Italia Nostra, a heritage organisation that went in for direct lobbying, and elite membership of interlocking establishment groups. The rescuing of Palladian villas was one of Italia Nostra's main causes. Other groups were devoted to landscape protection, zoning and the protection of historical monuments.

Italian politics in the 1960s was oriented towards economic growth, and the development of the countryside was not an issue of mass interest. In the 1970s, however, new groups arose that had virtually no links with the earlier groups devoted to the protection of nature. Mario Dani, historian of the Italian Greens, divided these groups into conventional, defensive and moderate, and unconventional, aggressive and radical (Dani, 1989). He describes how this original separation slowly gave way to a more unified sector, with a joint leadership. This merger was especially surprising at first sight because the radical environmental cause was taken up by the Italian Communist Party, who started an Environmental League (Lega per i L'Ambiente) in 1980.

The left had abandoned the radical movement of the 1960s and 1970s, the terrorist element of which had provoked public hostility, and adopted the so-called 'democratic road to socialism'. The new strategy was to gain legitimisation and normalisation for the Communist Party and other left groups, in order to stand against Italy's dominant Christian Democratic coalition. Radicals switched to communal action over causes such as worker health, which easily led to a commitment to environmentalism. As in Germany, they believed that transforming the capitalist social structure was an essential precondition for environmental preservation.

Disaffected activists also evolved a typically proto-ecological subculture, 'retreating', according to Dani, into 'neo-oriental religious groups, youth sub-cultures, anarchist and libertarian groups . . . anti-anthropocentric . . . anti-authoritarian' groups (Dani, 1989).

One of the first successful left environmentalist causes was the campaign against a chemical plant in Venice as a result of the Seveso accident in July 1976. One response was the founding in 1976 of a more scientifically oriented group, the Sapere Ecologia Medicina Democratica. As in Britain, the involvement of intellectuals who had a scientific background was crucial in giving the movement legitimacy and lobbying power.

The Seveso accident, which resulted in the release of poisonous dioxin, stimulated the formation of Milanese ecological groups, although neither the local population nor the more conservationist groups were mobilised. However, the government's 1977 plan for twenty nuclear power plants did produce joint action for the first time, with strong criticism from Italia Nostra and the left groups. Demonstrations, lobbying and requests for a referendum followed, and a Committee for Popular Control over Energy Policies was formed.

The Lega per i L'Ambiente, which was formally part of a cultural and leisure-time association jointly run by the Socialist Party and the Italian Communist Party, seemed quite independent of the latter. Indeed, the Lega attempted to ban hunting, a working-class and peasant sport in Italy, hence one supported by the Communist Party of Italy. The two wings of the ecological movement co-operated in demanding a referendum on hunting and nuclear power plants, a tendency that increased after the Chernobyl disaster.

In May 1985, when the Italian Green List stood in local elections, they won 600,000 votes, 2.1 per cent of the vote where they had candidates; 141 representatives were elected. The current situation is that the political activists have joined the moderates in tactics, and the moderates have joined the activists in issues, while the old working-class left defends its jobs and plants against the environmental demands of the new left and middle class – a phenomenon that appeared briefly in Germany.

Eco-terrorism in Italy appeared in the late 1980s in a particular political context, namely a passive and non-militant left and an extreme right-wing movement dedicated to terrorism, after a period of relative quiescence in violent or direct action.

Britain: the defusing of dissent

The British environmental movement has had an activist left demoralised by the non-revolutionary behaviour of the

Labour Party in power, but has also been affected by an element of Quaker and 'soft' Anglican support for a certain kind of left-liberal cause (overseas aid, refugees, other internationalist causes) into which British interpretations of Green politics fit very well. Many Green supporters were former Liberals, and funding bodies of Friends of the Earth, as well as Greenpeace, included the company that is the political branch of the Rowntree trusts.

So while Greens in Britain were linked to radical though politically respectable opposition groups who had accepted civil disobedience, the involvement of Quakers introduced a pacifist element that prevented violence becoming part of mainstream Green political strategy. Of course, the borderline between violent and peaceful civil disobedience can be narrow. The definition I would adopt is that peaceful civil disobedience places the onus of violence on the opposing party. The tactics of Greenpeace, who ram whalers and obstruct activities in circumstances that are bound to lead to violence, are a borderline case. However, the distinction becomes clear in the case of animal rights activism. This appears to be predominantly a British phenomenon. The movement's tactics started with relatively mild action such as releasing caged animals (the effect of releasing farmed mink into the countryside has been ecologically disastrous), raiding battery farms, and then breaking into laboratories to free animals used for vivisection. However, they soon escalated to attacking the persons and property of those involved in animal exploitation, even using bombs. Animal rights seems to be part of the anarchist underworld of British politics, and can be placed on the urban left of the political spectrum, together with the hunt saboteur movement. It is a specialised area of political action, and I will not consider it further in connection with ecological strategy, which is not necessarily associated with these activists.

The British Green belief in decentralisation goes with a belief in gradual reform and in individual activity, rather than organised mass action. The failure of such mass movements as the Campaign for Nuclear Disarmament (by failure in this context I mean the organisation's failure to achieve its main aim of unilateral nuclear disarmament) may have affected Green tactics, since so many Green Party members traditionally belonged to CND.

Why did national Green parties falter?

The fate of Europe's Green parties has been tied to that of Germany. Despite the untypical nature of the German Greens, they are the largest, the richest and up till now the most successful Green Party. Where Tasmanians and Americans alike save their rivers and preserve their wilderness, the German Greens at first glance seem to have lost sight of their original inspiration. Whether or not this is due to structural changes forced on them by party politics, the fact is that since 1980 the Green Party has described a parabola. In the first election after reunification (December 1990) support for the post-unification Greens dropped sharply, and although the Greens retained support in Hesse, there is little doubt that the underlying drive behind their party is diminishing. The former East German Greens actually won more votes when allied to the Christian Democrats.

Green parties have retained a role for the protest voter, especially in France and Belgium, countries much less affected by the German experience. Particularly in polls before elections, support for Greens can seem high. At the last moment, the support is not turned into votes. Why is this?

Opinion polls show the importance of the environment to the West German electorate. One reason why the other parties have fought off the Green challenge seems to be that they have embraced a modified version of conservationist and environmental rhetoric. But why should the interest in, and the urgency about Green politics not be transmitted into votes?

It is often argued that Green parties have not succeeded in Britain or the USA because these countries have two-party systems. Yet Green parties in countries with pro-portional representation have not done well. However, when Greens stood for the European Parliament in 1987 they attracted a substantial vote, some 15 per cent overall. This may mean that voters prefer to put their faith in supra-national organisations when it comes to issues that transcend national interest groups. Yet the European Parliament has very little authority, and no direct power. It cannot implement Green policies, only call for action from the Commission of the European Community.

In the last few years the media has taken up the cause of the environment. For TV programme-makers, the troubles are all man-made; the virtues all natural. The media may have helped to whip up support for Greens, but the nitty-gritty of Green policies is not terribly relevant to tender images of animals and wilderness. Those European ecological parties that evolved from agrarian and farmer parties have had very little coverage in Britain, while the involvement of media personalities (such as David Icke) has produced much more (it is true that cover-age of French ecological politics has been minimal, despite the charismatic personality of Brice Lalonde, but this kind of omission seems to be generally true of media treat-ment of French politics, and not confined to the Green side).

However, the gap between support for environmental issues and a vote for Green parties is wide everywhere in Europe. It is as if, despite the undoubted importance of Green issues to today's electorate, there is a reluctance to vote-in parties that put these issues first. Is this because Green parties are by and large committed to a range of non-specifically Green issues, even policies that contradict

Green thinking? The short-lived but impressive success of the German Greens would suggest the opposite: that environmental parties need to have a fund of left-liberal policies to prevent their being seen as excessively elitist, value-oriented, middle-class pressure groups.

Political scientists have puzzled over how to place the German Greens for some years, frequently ending by putting them with a cluster of alternative movements. In 1989 I came to the conclusion that the German Greens were not very Green at all (Bramwell, 1989) and the point has been made more recently in a study of Green philosophy (Dobson, 1990). It is the abandonment of 'Greenness' that accounts for the earlier success of the German Green Party. This may seem paradoxical, in view of the support for environmental policies in Germany. Yet it is a movement whose home is not that of party politics, because Green politics and institutionalised party politics are anathema, incompatible in process, in means and in ends.

Although the first electoral success was in Germany, the party fraction that succeeded was the urban left. Having hijacked the environmental train, the Greens briskly drove off in the opposite direction, attracting a scattering of disaffected civil service and youth votes in their wake. When the urban left lost credibility after the fall of East European communism, the German Greens declined too. And, as with the old joke about being in a rowing boat with an elephant (even a nice kind elephant is liable to tip it over), the other European Green parties were pulled down too. In the 1992 election the French ecologists suffered the same problem – French media interest, polls showing considerable support, but a drop in votes at the last moment, with protest votes going to other parties.

In the mean time, the real development of environmentalism was going on elsewhere, in the interstices of institutions, in pressure groups, and in the developing philosophy of the deep ecologists.

The global approach

The international approach to environmental reform fits in with the ideology of the radical ecologists, but also with that of the reformers. Supra-national control of human behaviour is an essential part of rationing resource use, and although ecologists stress decentralisation as a means to lower resource use, it is not clear how this would work in practice (see the earlier discussion of bio-regionalism). Even if bio-regions were created, they would not coincide with current political/national boundaries, and to put them into effect something more than a dissolution of such boundaries would be needed. One reason for the continuing faith in bio-regionalism is the Americo-centric approach of so many ecological theorists. Naturally, the USA could be divided into bio-regions without difficulty; but Europe and Africa would present problems. So in theory a strategy of capturing global institutions and using them to enforce ecological reforms might be acceptable to the radical ecologist.

However, so far in practice, such tactics have been used to further environmental reform in a legalistic and conventional manner. The internationalist views of the British Greens, in particular, have increased their tendency to look to supra-national organisations to implement their goals, just as liberal reformers did with UNESCO and the Food and Agriculture Organisation in earlier decades. The two agents most effective so far on the international scene are the EC and the United Nations. The case of the EC is especially interesting for analysing the ordering of environmentalist preferences.

One of the major points on the British Green agenda is physical and political decentralisation. It is argued that decisions should be made at local level. The reasoning behind this is not always stated explicitly. To some extent, it is part of an anti-urban and anti-megalopolis philosophy. The concentrating of people in towns, which are associated with industry and trade, is seen as anti-ecological, because it prevents people from developing a sense of responsibility for their own resource use and resource waste. In contrast, Greens connect decentralisation with local self-sufficiency, less trade, less movement of people, and less use of cars and lorries, with journeys made by public transport only where necessary. Although there is a strong Red-Green element in the Greens who espouse a more urban-oriented approach and reject what they see as sentimental and reactionary attitudes promoting rural life and the value of country living, they still support a 'community' approach to politics, which is decentralised although urban.

None of these aims is very obviously connected with the growing powers of the European Community. The EC is, or will be, internally a (relatively) free trade zone within a strong protectionist barrier, thus managing at one and the same time to offend two Green principles. One is the principle that trade with developing countries, although a bad thing, is better than unfair discrimination against such trade; the other is the principle that free trade, being a means to more trade between nations and regions, is also bad. The free movement of peoples, too, is against a half-hearted Green principle, which is not followed consistently (the basic position of British Greens is that the poetic, Celtic minorities in Britain, the Welsh, Scots and Irish, should be protected from contact with non-Celtic people, culture and habits, but that the mass movement of refugees into *England* should be encouraged as vigorously as possible). The EC also furthers the growth of multinational companies, by removing non-tariff barriers within its borders to the free movement of goods and services.

At the Green Party Conference in London in 1989, a motion to oppose British membership of the EC was put down for discussion. However, at the last moment the conference was told that the motion would be withdrawn. The reason for not wanting to air differences on the subject may have been that EC directives on pollution were likely to be more stringent and more carefully enforced than those imposed by the British government. Furthermore, they would, of course, affect all the countries within the EC. So to use the Community decision-making process, rather than to attack it, is a logical and effective way of tackling inadequate anti-pollution legislation. The Green Party was also beginning to realise the openness of European institutions (including its Parliament) to the Greens.

This signifies a top-level switch to reformist action within the Green Party. The rise of the European Green parties, and their success in the European parliamentary elections of 1987, was a further reason to support European institutions. The combination of European Green clout and an institution where policy-making is largely carried out by the Commission, and then rapidly rubber-stamped by the Council, meant that British Greens could, at one jump, find themselves in a position to do some effective lobbying that would have taken years if not decades within Britain itself.

So talk of decentralisation was put on the back burner where EC directives on environmental matters were concerned, and the prospect of an EC environmental agency with something approaching the powers of the American Environmental Protection Agency has been welcomed by active Greens in Britain and Europe. In turn, the funds available for research and studies into environmental pollution have been as naturally courted.

In effect, therefore, the relevant directorates-general of the Commission of the EC have provided a parallel political platform for Green lobbyists, one which is more effective than the equivalent British one, and which moreover possesses the power to set and enforce a common level of environmental protection and anti-pollution laws in the countries where its writ runs. What was on offer was the prospect of rapid legislative action on pollution problems, or at the very least an acknowledgement that such problems existed and should be tackled at a supra-national level. It is not surprising that this prospect should be more attractive to Green activists involved in party politics than continuous engagement in the apparently never-ending and fruitless Green Party infighting.

The reformist-left quality of British Greens can be seen most clearly in their approach to the EC. It is a body capable of massive regulatory intervention, the method most favoured by the reformist left, and seen by the liberal as a means of creating a framework within which a market economy can function. So when British Greens oppose the poll tax for being undemocratic and for being the product of central government, it is understandable that they should nonetheless support taxation that goes to support the central and much less accountable decisions of the EC.

Global agencies such as the UN are seen as an effective strategy for *environmental* issues, rather than ecological ones: for example land management, wildlife protection, nature reserves, deforestation, soil and air pollution and issues concerning the sea are all currently the subject of conventions and laws emanating from international agencies. Whereas the economic problems of the developing world have proved intractable to action by international aid agencies, the physical environmental problems – those which cut across national boundaries and need international agreement to solve – seem suitable cases for treatment, equivalent to the efforts by the World Health Organisation to eradicate smallpox, or the Food and Agriculture Organisation's attempts to increase food production by tackling insect infestation.

This is still a far cry from the original hopes and ambitions expressed when the United Nations was founded. Sir George Stapledon, founder of grass management science, placed his faith in the UN as a means of reorganising human life on the planet. In *Human Ecology* (1964, written in 1944) he described the UN as a means to plan human societies more ecologically, with particular reference to efficient resource use. However, the UN's foundation coincided with a peak of optimism about the practicalities and virtues of modernisation. Environmental agencies were created, but the development aims of the UN vitiated their usefulness. These bodies remained relatively power-less until the late 1960s.

Several factors contributed to their renewal. If sustain-able development were to be taken seriously, consensus on changing patterns of consumption, production and trade could only be developed at this level, and the United Na-tions Commission on Sustainable Development has been crucial to this work.

The role of the Greens in the European Parliament has remained ambiguous. Although the Maastricht Treaty gave the European Parliament more authority over environ-ment (among other sectors), the 1994 election results did not show an increase in the Green vote (Luxembourg and Ireland excepted): on the contrary, their seats fell from twenty–seven to twenty–one,* and Green representation did worse in many countries than this figure would suggest, down heavily in Denmark, Italy, France, Belgium and the Netherlands. However, this election seems to have reflected national politics for many countries involved, rather than demonstrating a 'European' viewpoint: and thus the results shadow the secular decline of the Green parties in national elections. The exception was Germany, where the Greens indeed seem to have replaced the FDP as minority coalition partner in Germany, having developed a range of fairly conventional 'nice' left policies and abandoned their funda-mentalist ideas.

Given the change in the position of the European Parlia-ment the next European election should show a develop-ment towards a genuinely multi-national European set of politics. If so, then the role of the European Commission in implementing the sustainable development Action Plan will be crucial, because it is the only multi-national bloc committed to sustainable development at all. And yet, do we even know what it is?

* *The Economist*, 18 June 1994, p.44.

CHAPTER SEVEN

Sustainable Development: Questions, not Answers

THERE HAS BEEN a speedy acceptance of the concept of 'sustainable development' over the past few years. It is enshrined as a policy aim in the European Commission's Fifth Environment Action Plan (1992). Governments have departments dedicated to promoting it. Top businessmen have joined forces to promulgate it. Indeed, sustainable development has become one of the key phrases in policy-making in the 1990s. Yet among those engaged in the subject, there appears to be little consensus on what the phrase includes. Writers have traced literally dozens of different uses of 'sustainable development' (OECD, 1989, p. 3).

In part this is because 'sustainable' was a word that became attached to a variety of policy areas during the 1980s, and different definitions may simply reflect functional differences. But discussions about the meaning of 'sustainable development' also reflect real differences in policy approaches, particularly concerning the achievability of sustainable development: they thus relate to key issues and policy initiatives.

The term first appeared in 1980, when the International Union of Nature Conservation referred to 'sustainable development through the conservation of living resources'. The best-known definition comes from *Our Common Future*, the report by the Brundtland Commission:

Sustainable development is development that meets the needs of the present without compromising the ability of future generations to meet their own needs. (World Commission, 1987, p. 43)

Environmental economists later in the 1980s defined 'sustainable development' as a way of 'being fair to the future', meaning that we need to leave to future generations the environmental resources we inherited. In this form, 'sustainable development' stresses the need to conserve finite resources and prevent environmental degradation.

The Brundtland definition, however, continues:

> It contains within it two key concepts: the concept of 'needs', in particular the essential needs of the world's poor, to which overriding priority should be given; and the idea of limitations imposed by the state of technology and social organisation on the environment's ability to meet present and future needs. (Ibid., p. 43)

While environmental economics already in principle depended to a degree on global policy-making (because environmental problems are transnational, but also because solving them imposes economic costs that can hamper international competitiveness), the Brundtland Commission, by linking the ideology of development with the idea of sustainability, set in motion a change in global policy-making by aid and donor agencies, international groups and non-governmental organisations.

Turning the 'buzz-word' into tangible policies, working on its practical applications, brought new shifts of emphasis, particularly where the 'development' part was concerned. Sustainable development implies that sustainable growth is possible, whereas to the deep Green growth will always lead to environmental damage. In this most Green of concepts, therefore, lay the operational destruction of deep Green ideals, by 'normalising' the growth ethic, by accepting the goal of maximising human welfare.

Before the involvement of international development agencies, in the late 1970s and early 1980s the emphasis was on sustainable criteria in industry, that is, were the resources used renewable? Some preferred to use the phrase 'sustainable use of the environment', because it is a more precise definition, implying the substitution of renewable resources for non-renewables.

Interest in sustainable development was stimulated in the 1980s by the increasing refinement of environmental economics, a discipline whose best-known exponent was David Pearce, Professor of Economics at University College, London, and co-author of *Blueprint for a Green Economy*. He defined sustainable economic growth as follows:

> Sustainable economic growth means that real GNP per capita is increasing over time and the increase is not threatened by 'feedback' from either biophysical impacts (pollution, resource problems) or from social impacts (social disruption). (p. 33)

Environmental resources were added to the factors necessary for sustainable welfare. Sustainable development, then, meant valuing finite resources with a view to their value to future generations; it meant valuing the contribution the environment made to human well-being, and, finally and most problematically, it meant valuing qualitative aspects of the environment in aesthetic and spiritual terms. But in order to cost resources, we had to know what they were; what exactly was the rate of species depletion, for example. If we were to look at growth in terms of environmental capital, we had to know the 'quantity and quality of the natural resources base of a region' (Kruik and Verbruggen, 1991, p. 2). As part of evaluation, environmental ethics suggested that an ecocentric point of view should allot 'rights' to natural objects (so far, to trees, rivers and some mammals).

The Brundtland Report merged the ideas current in environmental economics with the problem of global development. How were developing countries to be enabled to grow without contributing to pollution? Was it essential to have low growth or no growth in the developed world to compensate for the pollution caused by growth in the developing world? Or could technology transfer and other policies lead to sustainable (i.e. non-polluting) growth? Furthermore, was growth itself the right aim? If we had indeed reached our 'limits to growth', then further growth should be sacrificed to the well-being of the planet. But such a decision was perceived as unfair to the poor and to developing countries which had not yet reached a satisfactory state of growth.

These arguments were summarised by Robert Goodland in a 1989 paper for the World Bank.

> We must shift our goal from quantitative growth to qualitative improvement, or development defined in qualitative terms, to enhance the enjoyment we all get from a sustainable level of production. Sustainability is not yet operationally defined ... because the concept is not simple, definitional approaches rapidly become politically sensitive and expose messy unresolved issues. ... Sustainable growth and sustainable development are not synonymous. (Goodland, 1989, p. 10)

The argument that there had to be natural limits to growth rested not on models, which had been proved inadequate (Meadows *et al.*, 1972), but on the belief that growth was a function of the environment in two ways: the 'source and sink functions'. Resources came from the environment, and waste went back in it. These inputs and outputs were natural limits to finite resource use and to waste assimilation.

Global equity, however, was a politically sensitive issue. If use of finite resources was to be reduced and, eventually, halted, then there was a cost involved: the cost of resource substitution, and the cost of a lower standard of living in some ways if resource substitution proved impossible. Who should bear this cost? If developed countries were to be persuaded or bribed into environmentally acceptable resource management, who should take on the responsibility, and how was it to be done? Who was to organise, to distribute, to ration and to police this system? Environmental economists believed that one way out was to develop ways of costing resources, both finite and renewable, which would be more 'real' than market valuations. If policymakers within governments had this as a tool, then they could make decisions about development and growth that would be more resource efficient and perceived as more adjustable than current ways of measuring economic growth.

Given that it should be possible to know what was going on in the national environment – trees cut, fish fished, land used – what was the ideal rate of use? Where determining the saturation point of soil, air or water was a simple measuring exercise, 'rates of use of renewable resources' was perceived as a normative issue. Struggling with this question, Kruik and Verbruggen wrote:

> To answer the question whether the development of a region or a nation is sustainable or not, we must first decide which developments are important for the sustainability issue. . . . This question has scientific as well as ethical angles. Do we just look at the sustainability conditions for the human species, or do we include other species as well? (Kruik and Verbruggen, 1991, p. 3)

Environmentalism in the 1970s and 1980s pursued the aim of maintaining ecosystems, preserving bio-diversity and not overloading the carrying capacity of the earth. Economic growth seemed to many environmentalists to conflict with these aims, because it entailed the use of finite resources, despoliation of wilderness and the growth of waste by-products. Others wondered if there was a way of reconciling economic well-being and maintenance of the biosphere. The International Union for the Conservation of Nature, in *World Conservation Strategy* (1980), expressed this aim as

> Sustainable development through the conservation of living resources. . . . The maintenance of essential ecological processes and life-support systems, the preservation of genetic diversity, and the sustainable utilisation of species and eco-systems. (International Union, 1980)

The attempt to merge the goal of sustainable resource use with that of global human well-being produced the Brundtland definition, enshrined in the proposed legal principles 'for environmental protection and sustainable development' (World Commission, 1987, Annex 1, p. 348):

> All human beings have the fundamental right to an environment adequate for their health and well-being.
> States shall conserve and use the environment and natural resources for the benefit of present and future generations.
> States shall maintain eco-systems and ecological processes essential for the functioning of the biosphere, shall preserve biological diversity, and shall observe the principle of optimum sustainable yield in the use of living natural resources and eco-systems.

Sustainable development, then, has four main principles. Two of these are 'strong' definitions – that there shall be no further depletion of natural resources, and that global equity should be a factor in balancing human welfare against maintenance of a stable ecosystem.

The other two principles are 'weaker' ones – that renewable resources shall be substituted for finite resource use, and that national growth policies should use information gained for monitoring the environment's natural resource base and absorption capacities. These principles are affected by, but not dependent on, contemporary fears about the greenhouse effect, and other impending global crises. These two weak principles can also be seen as tools for achieving the larger goals.

Sustainable use of renewable resources has to satisfy various criteria. One is that use and disposal of the resource has to stay within ecological limits: that is, the resource must not be depleted so much that it or its related ecosystems or species are in danger of extinction. Another is that the resource must not be polluted beyond a certain 'sink' capacity. A third is that the resource (or its equivalent) must be renewed as far as is possible.

The Environment Act of New Zealand (1986) sets sustainable management of natural resources as an objective with a Resource Management Law which lays down procedures to institute sustainable management. Under the Environment Act, the Ministry for the Environment provides environmental advice to Cabinet, and can refer the sustainability objective of the Environment Act to government decision-makers. This is an example of giving environmental claims an overall say in policy-making, one of the key issues mentioned below. Natural resource accounting is being developed as a tool for renewable resource management.

The earth's 'carrying capacity' was defined in *Global 2000* as the ability of biological systems to provide resources for human needs. This concept has been criticised for being too anthropocentric, and for being too vague, since the earth's capacity to support human needs is hard to calculate, depending as it does on technological advances and social, economic and institutional arrangements. However, a start has been made in analysing how far ecosystems and resources can function under stress; and investigating how much pollution and waste the ecosystem can absorb and process without degradation.

The Dutch report *Concern for Tomorrow* (RIUM, 1991) calculated maximum permissible per capita figures for polluting emissions, and their National Environment Plan Plus targets for reduction of emissions, waste and resource use are in part based on these calculations. An example of these maxima is the calculation that each person can produce 6 tonnes of carbon dioxide per year. According to this calculation, the developed world is over its 'ration', while the developing countries still have space to go up. Relative energy efficiency and productivity is not taken into account by this emission 'quota'.

The critical load approach to environmental pollution was developed in the course of dealing with the problem of acid rain in soil and water. It is based on the assumption that the ecosystem can absorb a certain amount of pollution, and that under a certain level or threshold, the pollution does not matter. Critical load is used for ecosystems where there is a buffering capacity: it defines the tolerance levels needed for the ecosystem to stay in a steady state. For soil and water pollution, the concept of critical load has largely replaced the concept of a 'dose response' relationship, that is, a linear-type one-to-one relationship between pollution and damage. It can, however, be established that over a certain pollution limit soil and water become unable to sustain life adequately. In some areas, such as human health or damage to stonework, critical load criteria are not applied; the dose response approach is used.

The critical load approach is a threshold approach. Where a target load is decided on by policy-makers, it may be set higher than the damage threshold, for cost or other practical reasons. It should be noted that in Scandinavia some scientists say that the acid rain tolerance level of the soil is zero, that is, that the system has no buffering capacity.

Given sustainable development's emphasis on leaving what we inherited to future generations, ways of managing finite resources are crucial for putting the idea into effect, and particularly problematical, because finite resource management strategies are at a very preliminary stage of development. According to orthodox economics, when a resource becomes scarce, the price goes up, thus encouraging substitution by another resource. The problems with this free market approach from the environmentalist-economist point of view include the fact that the resource may become very scarce, almost used up, before the price-rise effect begins to work. This means depriving future generations of most of the resource. Another is that the substitute material may produce environmental problems which will not be obvious (CFCs are an example of this). It is argued that the use of the price mechanism unfairly penalises the poor and developing countries. Lastly, there is no obvious substitute for some environmental goods (such as the ozone layer, or natural atmospheric regulatory systems).

The long-term aim of finite resource management is to stop using the finite resource. This is impossible to do at once because of the economic dislocation that would result. The problem is how to decide on an appropriate discount rate to value finite resources according to their scarcity. One suggestion is to divide finite resources into strategic ones – which are essential, vulnerable, non-substitutable and scarce – and the others. Another idea is to look at the 'waste sink' angle: if resource use in production is reduced, then waste and emissions are reduced, enabling economic development to continue without using up the environment's buffering capacity.

Waste minimisation strategies are being adopted by Europe and the USA because of waste disposal problems, such as shortage of landfill sites and dangers to human health and the environment caused by toxic waste. Reducing waste, both toxic and non-toxic, is crucial to a sustainable development policy.

Waste minimisation involves reducing waste either at point of source, or at point of disposal. Point-of-source reduction includes reduction in packaging; point of disposal reduction includes recycling. Recent legislation in Germany affecting the car industry is an example of an industry becoming subject to waste minimisation regulation. Car manufacturers are now responsible for the recycling of their own cars; the manufacturers are compiling a labelling system for plastics used and are redesigning their cars to make recycling easier. German recycling policy is more of a trade war weapon than a serious contribution to reducing wastes; most recycling policies rapidly produce too much of the recycled material, and are not cost-effective. Nonetheless, perhaps because useless packaging is something that everyone perceives every day of their lives, recycling waste is a popular policy. Composting or burning household waste is a policy being examined by some countries, and this probably has more relevance to the problem.

Clean technology is another term with definitional problems. Clean technologies focus on elimination of pollution problems at source. The term 'clean process technology' embodies the aim of cleaner production within the factory. There are three approaches to clean technology: the introduction of new processes or products to prevent or reduce the production of dangerous or unpleasant by-products; modification of existing production processes that would otherwise produce pollution; and cleaning up existing pollution by reusing waste or recycling it back into the production process.

Clean process production would involve the following type of measure: changing materials to less polluting ones, for example not using organic solvents, recycling materials within the production process (e.g. cleaning and re-using water used for discharges), and adjusting design to avoid unnecessary processing, for example not setting lathes to grind to unnecessarily strict standards.

Typically, in industrial countries chemical companies produce 20 to 70 per cent of all hazardous waste, either through manufacturing, or through the final product. For this reason, according to Frances Cairncross, the most radical corporate thinking on the environment is taking place in European and North American chemical giants (Cairncross, 1991, p. 183). However, although companies like Monsanto and Dow aim to be ahead of environmental regulations, they regard the aim of zero emissions as unattainable. In other words, they do not see sustainability as a realistic aim.

The Brundtland Commission guideline is that we should leave the next generation a stock of assets no less than those we have inherited. Inter-generational equity is a powerful intuitive ideal underlying sustainable development. As David Pearce says, 'we can meet our obligations to be fair to the next generation by leaving them an inheritance of wealth no less than we inherited' (Pearce *et al.*, 1989, p. 35). The need for rules or principles of management is inherent to the inter-generational aspects of sustainable development. The danger of irreversible environmental loss or damage and uncertainty about future needs requires us to maintain a 'safe minimum standard of natural capital', or to establish a 'precautionary principle', according to which the overall stock of environmental resources should not be allowed to diminish over time (OECD, 1989).

Leaving the existing stock of natural capital to the future generations means making assumptions about their future needs. It has been argued that we cannot do this, and that it is not necessary to 'second-guess the future'. Others have pointed out that it is merely prudent to assume that the next generation will need air, soil, water, forests, fisheries, bio-diversity, energy and minerals as much as we do. But how to ration resources between present and future generations, when we cannot know what kind of societies will exist, nor what their resource needs may be, nor how many generations we are to allow for, presents obvious difficulties. If each generation looks after its obligation to the next one, no single generation needs to worry about those far into the future. However, as Pearce also points out, in practice, environmental problems – such as radioactive waste and species loss – can persist for many generations.

The concrete interpretation of this ideal is that finite resource use should drop to zero as soon as possible. Pursuing this ideal, the Dutch Ministry of the Environment is anxious to prevent further exploration for minerals. One recent suggested target is to aim to halve finite resource use each year, thus allowing time to shift to renewable resources. A more developed idea is to harvest renewable resources on a sustained yield basis rather than 'mining' them to extinction (Goodland, 1989, p. 14).

Non-renewable resource extraction, however, can never be sustainable. The World Bank suggests:

> Whenever a project cannot be sustainable, sustainability can be attained only by investment in a complementary project that would ensure sustainability for the two projects taken together. An example could be an iron ore mine surrounded by tree plantations. A portion of the receipts from the mining should be invested in a renewable asset annually. The growth rate of the renewable asset . . . provides a permanent income stream to be adjusted to the depletion of the finite resource. (Ibid., p. 15–16)

Hermann Daly suggests that the benefits of sustainable projects should not be calculated in the same way as for unsustainable ones. Discount rates that reflect rates of return on unsustainable uses of capital do not reflect the true 'cost' of the project. His example is of a sustainable managed forest, yielding 4 per cent, as opposed to clear felling of a forest, yielding 6 per cent. The sustainable project should be preferred.

Choosing projects with a low discount rate instead of a high one may not appeal to investors, unless it is presented as a long-term investment. Presumably, recommendation to invest on a low-discount-rate basis is meant to apply to aid and donor agencies, and to international financial institutions, such as the European Bank for Reconstruction and Development or the European Investment Bank.

Environmental economists disagree about whether economic growth can be compatible with sustainable development. It is suggested that if growth is redefined to include environmental and quality of life factors (health or longevity, for example), then growth as an indicator becomes less important. Aside from the theoretical problem of working towards a zero growth, steady-state economy, the debate covers the problem of the demand for economic growth coming from the developed countries. Given the resource management and pollution constraints of the sustainable development ideal, this demand can only be satisfied by balancing positive growth in developing countries with zero or negative growth in developed countries.

The wealth of a nation may be divided into its human capital (the skills, education, innovatory capacities and social institutions), its stock of man-made assets, and its natural capital. Natural capital includes such environmental goods as breathable air, clean water, unpolluted and uneroded soil, and waste sink capacity. It also includes finite and renewable assets such as forests, mines and fisheries.

Sustainable development seeks to avoid irreversible

changes (or losses). Irreversibility is considered by proponents of sustainable development as more likely to occur with natural capital, since although renewable resources can, by definition, be renewed, many natural phenomena and systems cannot: for example primary forests, species, and complex ecosystems such as certain wetlands. Much of man-made capital, by contrast, is perceived by them as renewable, although, again, much is not (historically and aesthetically valuable artefacts, for example; one might argue, too, that human capital is hard to build up once destroyed). Thus, the substitution of natural capital by man-made wealth is not considered 'sustainable'.

Natural resource accounting is one of the tools used to monitor progress towards sustainable development. It is intended to correct the omissions of the standard system of national accounts (GNP), which records the stocks and flows of an economy with a view to comparing economic growth rates. Standard national accounts, which are based on Keynesian concepts of aggregate demand and growth, have been criticised for ignoring qualitative aspects of growth and of well-being, and for ignoring hard-to-quantify economic inputs, such as housework, and, in countries with a substantial peasant population, self-sufficient consumption. National accounts do not allow for the depletion of natural resource stocks – extraction of these is counted as income. Pollution control costs appear as an increase in the national product, which is misleading. Natural resource accounting is an attempt to evolve a system of national accounts to correct these deficiencies. It should also help to monitor the degree of sustainability within an economy and indicate gaps in our knowledge of the environment.

The World Bank and United Nations Environment Programme (UNEP) have developed a framework for 'satellite accounts', using inputs from UNEP on documenting natural resources. Various countries have initiated natural resource accounts (NRAs), using different approaches. They include the following:

Norway: a physical inventory of forest, soil and water resources together with 'emission accounts', added to the 'material accounts'. In place, but not considered to be a complete NRA, because it does not value stock changes.

France: the 'Patrimoine' system, a regional inventory of physical units, not yet complete.

Canada: the Institute for Research on Environment and Economy, Ottawa, has been active in developing the concept of 'The State of the Environment' (SoE) report. Unlike natural resource databases, SoE looks at 'qualitative states' of the environment in different regions, and does not attempt to aggregate the data. They distinguish between 'biological accounts', 'geological accounts' and 'cycling systems accounts' (e.g. soil creation and erosion, effects of earthquakes and volcanoes).

The USA and Japan: correction of GDP through monetary valuation of environmental damage.

Indonesia: the World Resources Institute published an NRA for Indonesia in 1989 (Repetto, 1989). It aimed to adjust GDP figures for natural resources stocks by taking extraction rates of forest, oil and soil (depletion through crop production and erosion) into account. The cost of environmental damage was not included. When adjusted by these factors, average annual GDP was 4 per cent instead of 7.1 per cent.

At present, the lack of uniformity in systems of natural resource accounting, and the as yet unsolved problems of valuing resources and pollution damage, mean that NRA can only help monitor some aspects of sustainable development (Arntzen and Gilbert, in Kruik and Verbruggen, 1991, pp. 45–55). Hans Opschoor has pointed out that NRA fails to include transnational pollution and resource use (in Kruik and Verbruggen, 1991, p. 10). He considers that the effect of a country's 'total environment pressure' should also be monitored; but it should be regarded as a separate exercise from monitoring a country's environmental capital.

According to Hans Opschoor and Lucas Reijners, 'sustainability indicators reflect the reproducibility of the way a given society utilises its environment' (ibid., p. 7). Indicators have to be developed in order to measure how sustainable a nation's economy is. Sustainability indicators are not intended to be snapshots of the environmental situation, but to be normative: that is, they should show how far the situation is from a desirable norm, or from a reference situation, either in the past or in the present. Thus, like the critical load concept, indicators must relate to policy-determined emission maxima, and a policy-determined growth path.

Indicators are needed to cover three areas: pollution, resources and biological diversity. Examples of indicators in a sustainable economy (defined by Opschoor as a steady-state economy) would include the following:

Use of renewable resources should not exceed the formation of new stocks.

Use of relatively rare non-renewable resources should be close to zero, unless substitution for future generations is achieved.

Use of fossil fuel would be permitted if equivalent biomass of solar energy devices were put aside for future generations.

Use of relatively abundant non-renewable resources is permitted if increasing expense of extraction is compensated for.

Accumulated pollution in air, sea or soil is not permitted.

This steady-state definition of sustainability is disputed by other writers (Braat, in Kruik and Verbruggen, 1991, p. 62), who point out that natural communities and ecosystems are highly dynamic. Models of future studies show that there are different patterns of development of the major variables of man–environment systems that can be called 'sustainable'. Braat concludes from this that sustainability indicators need to be defined in terms of development curves, which can be compared to acceptable patterns of development (ibid., p. 64).

A different approach to the search for indicators is found in Canada, where the Canadian State of the Environment team are looking for indicators resulting from the monitoring of the health of ecosystems, the capacity of the system to regenerate, its stress level, the impact on human health of environmental degradation, and the rate of extraction of finite resources. The statistics 'indicate' sustainability levels. Whether sustainable development is compatible with growth is a hotly debated issue. Some believe that by substituting renewable materials and reducing waste streams and emissions, 'clean' growth is possible; others, including the New Economics Foundation, the Centre for our Common Future in Geneva and personnel in the Dutch Department of Statistics, believe that 'growth' is a misconceived ideal, best abandoned as soon as possible.

On the knotty problem of the political acceptability of no-growth policies, attitudes vary. Some argue that as sustainable development is accepted by all world leaders and spokesmen as a goal, it must represent popular opinion. Others point to the eager consumer acceptance of recycling and pursuit of Green products as a sign that public opinion would be behind no-growth policies, providing they were explained clearly and a crisis atmosphere engendered (for example the greenhouse effect, the ozone layer). It has been argued that for a country to introduce sustainable development entails a policy of no trade. This is because it is impossible to be certain that imported goods have been produced under conditions of sustainability. Others argue that we should continue importing goods from developing countries but improve their terms of trade and alleviate their debt burden.

The Institute of World Economics at Kiel sees material balance accounting as the next step in sustainable development policies. So far, national pollution control policies tend to be oriented towards specific emission sources. An accounting framework is needed to cope with high-aggregate, low-point source-polluting emissions. Kiel would like to research into a legal framework that allocates strict liability (and throws the balance of proof of safety on the user). For example, cadmium, a dangerous heavy metal, could be used, but the user would have to guarantee its safety.

The Centre for our Common Future in Geneva and the World Resources Institute in Washington, DC both see the use of local technology and knowledge in developing countries as a key issue in sustainable development. The World Resources Institute believes that environmentalists must be given a major say in planning at government level. The problem is to avoid having a marginalised and weak Ministry of the Environment, which has no control over the planning decisions of other ministries. This would mean either appointing environmental assessors to each government department, or having a Ministry of the Environment which advises each department (as in New Zealand). A strong management team and clear targets are pre-requisites for instituting an environmental plan.

Sustainable development started off as a global concept, but in practice is being developed at micro-level, by local authorities and municipalities, since governments cannot grasp the nettle of zero growth of population control. However, a council can monitor its waste disposal, sewage and water policies, can sometimes control energy prices, and is usually in charge of land use.

The furthest advanced in environmental management is the Dutch government, which has set specific targets and dates for reductions in polluting emissions. It hopes to achieve this reduction by consensus, but is prepared to move to coercion if the consensual approach fails. However, it has not yet produced a policy for resource use within the Netherlands. Creating an integrated programme of this kind requires good data and a consensual social structure. It is too early to say that it cannot work, but the jury is still out. It may be that the claims of the deep Greens that a much more fundamental change is needed will be proved correct – accepting for the sake of the argument their parameters – because if the West cannot go Green without pain, it will not do so. And if it will not, no one else will. And where do Greens go if consensus fails?

CHAPTER EIGHT

Deep Ecology and Civil Disobedience

ONE OF THE most intriguing developments in Green circles has been the growth of the Green saboteur. This chapter examines the doctrinal roots of Green civil disobedience. While the rise in both civil and uncivil disobedience can be partly attributed to the disillusionment of the left with former orthodoxies, it can more helpfully be attributed to the ideology that has inspired them. Environmentalism is a radical belief with a hidden history. It is not in the same position as other ideologies such as socialism, which, as Andrew Dobson has perceptively suggested, was founded on liberal claims about equality and liberty, and fed on liberalism's 'apparent failure' to make its own ideals work (Dobson, 1990, pp. 7–8). Direct action groups are a predictable consequence of a religious ideology. Deep ecology is such an ideology. It forms the radical edge of the Green movement, and the most consistently anti-humanist trend within it.

The deep ecologist activist resembles an angry and disenfranchised child. The search for supra-rational modes of sensory enhancement through communion with the earth is an attempt to escape from human limitations, to reach super- and supra-human status. The metaphors of earth mother and father rapist that permeate deep ecologist literature are Oedipal; they show the urge to kill the father and marry the mother. The aim is to merge with the mother earth, the tactic to use her strength and skills to destroy the interloper.

Deep ecology is an approach to environmental issues first formulated by Arne Naess in 1972, but implicit in ecologism from its inception. It is apocalyptic, anti-political and anti-reform. It has adopted principles of biological equality, and emphasises the role of humanity as a participant in nature rather than as nature's steward. Hence its anti-humanism. The deep ecology vision of humanity is as a natural disaster, something like an exterminatory virus.

From this belief, several forms of action flow. The assumption that nature is benevolent and harmonious leads to the belief that it is possible to solve conflicts and problems without harm to any species, simply by giving up unnecessary and extravagant ways of living, and adopting a more sensitive approach. When explored, however, this apparently innocuous picture turns out to mean living at a Third World level; not the affluent Third World, either.

> If life under these circumstances sounds like reproducing the styles of life most often associated with Third World countries, then the Green position on trade (and not a few of their other recommendations) reflects Rudolf Bahro's view that 'With a pinch of salt one might say . . . the path of reconciliation with the Third World might consist in our becoming Third World ourselves.' (Dobson, 1990, pp. 106–7)

It means protectionism, giving up what the deep ecologist believes to be unnecessary fripperies, restricting personal mobility, and introducing the authoritarian structure of a command economy in order to allocate resources. Which lifestyles are praised by ecologists? They admire the closeness to the earth of the American Indians in the nineteenth century, and of the Amazonian Indians now. It is a mistake to see this admiration as a variant of the Noble Savage ideal or fantasy. Deep ecologists really would like to introduce these lifestyles. And since the broad masses, in either hemisphere, are unlikely to want to do this without coercion, and population reduction by one means or another is a crucial part of reducing resource consumption to an acceptable level, the problem for the deep ecologist is how to use pacific means to make people change, and how far violent means are justified.

Those deep ecologists who believe in the inevitable coming destruction of the ecosystem feel that man has only himself to blame, and that the earth can and will shrug him off and take care of herself. They, the saved, know that the survival of nature is more important than that of humanity. Others believe that persuasion and education can be carried out in time. Where both disagree with reform ecologists is in their rejection of political parties and the possibility of institutional change. Although they are sometimes prepared to throw their weight behind single-issue environmentalists, deep ecologists are not happy with technical solutions to environmental problems (Postrel, 1990, p. 2; Dobson, 1990, pp. 138–40), because recycling or reducing pollution loads through technological change does not *reduce* consumption and resource use. There is considerable anecdotal evidence from the USA of cases where solutions to environmental problems have been attacked by ecologists, because they might lead to the continuation of what they see as a profoundly wrong way of life. To them, reforming the system means accepting inauthenticity – a reaction emanating from the wilderness-inspired values of American environmentalism (Postrel, 1990).

An important preliminary to understanding deep ecology is to see what kinds of minds have been attracted by its ethic. Would one expect them to include the scientists against science referred to in earlier chapters? Many of the converts to ecologism since the mid-1970s are not themselves mystics, but began by deploring mysticism, occultism and the abandonment of a belief in science, and were later converted to a resource-egalitarian, Doomsday ecologism. Examples in this category include Peter Medawar, Carl Sagan and Isaac Azimov. All three are (or were) distinguished scientists, and have written critiques attacking pseudo-science, Danikenien theories, and mystical cosmologies. However, all three took up the resource egalitarianism that lies behind recent Green Doomsday thinking. Clearly, when Greenness is presented in a certain kind of way, it convinces.

The justification by science and scientists has to be incorporated into a history of ecological thinking if we are to understand its force, its convincing nature, and its rounded and comprehensively satisfying ideological completeness. In order to be forced into taking direct action, anyone other than a thug has to be absolutely sure that they are right in their beliefs, and the moral justification of direct action to deep ecologists lies in this certainty.

But where does the moral backing for deep ecology lie? Not in human values, since for ecologists these are perceived as subjective and unreliable *Homo-sapiens*-centred frauds. The human value system cannot be trusted, because humanity is flawed. Its tendency to destabilise and destroy the ecosystem, nature or even the earth itself is inherent in the spiritual flaws that developed along with civilisation, agriculture, technology and religion. So the justification for the deep ecologists' programme of action has to lie outside the human value system. In this, deep ecologism has the power of religion, which claims otherworldly revelation. But ecologism does not reject the world. On the contrary, it celebrates it. So the values have to lie in the world, in nature itself. The natural system has rights. Deep ecologists have to believe that they can leap over inherent human flaws and latch on to nature's way of being, fit into the harmony of the universe. Deep ecologists stress meditation in natural surroundings as a method of sinking into the rhythms of the earth, and coming intuitively to a doctrine acceptable to nature. Nature itself provides the values, but these are transmitted according to the deep ecologists' perception of them. Despite the apparent humility of the ecologists' stance before Nature, they are people who have access to special knowledge, who become part of the elite: who are saved.

The second justification for direct action lies in the belief that deep ecology is backed by a scientific assessment of what is going on. In view of the deep ecologists' many attacks on science, this may seem a strange statement, but ecologism in general is characterised by the paradox that ecologists perceive science as bad, yet rely on it. In reality 'science' is a bundle of related disciplines, not a unity; however, it is a truism for Greens that 'science' equates with the mechanistic world-view they believe was created by Descartes, Newton and Bacon (Rifkin with Howard, 1980; Merchant, 1980). For a long time ecologists and environmentalists have been scientists. Indeed, the essence of ecologism has always been propagated by a small section of the trained intelligentsia. Not just because the first political ecologists usually *were* ecologists, plant scientists, geographers and geologists, but because it takes a rational and trained mind, expert in its own discipline, to have the confidence to assume that human beings can diagnose such large-scale syndromes of global sickness and can then cure that sickness. The tendency to credulity typical of deep ecologists arises precisely because they believe they can rely on science.

The commonsense view of science and the scientific method is that it is the province of reason, of hypotheses neatly made, evidence produced, arguments proven or disproven. This is a summary of Sir Karl Popper's argument, and Popper was extremely hostile to anything smacking of 'holism' or, even worse, vitalism (Popper, 1982, p. 137).

Yet in reality, science has always walked hand in hand with ideas that do not fit that picture. The counter-intuitive nature of science was, in 1993, the subject of a fascinating book by Lewis Wolpert. Perhaps because of this, scientists have to be prepared to reject 'commonsense beliefs'. Newton believed in alchemy and Rosicrucianism. The scientific efflorescence of the Renaissance was connected with the creative cosmologies of Neoplatonism, not the rational discourse of Aristotle. The German philosopher Schopenhauer commented admiringly in the nineteenth century on the excitement and fantasy involved in the idea of that mystical and invisible substance – electricity (Driesch, 1914, p. 121). The world of magnetic fields fired scientific thought for a few decades, and still underlies the work of Rudolph Steiner. Astrologers believe in them too; astronomers are just beginning to return to them. Yes, the wonders of science are manifold. It takes a creative mind to invent quantum physics or to believe in relativity, and it takes a certain kind of credulity to take these wonders on trust.

Another characteristic of the scientific mind is to believe that once successful in one field, you can solve problems in others. Scientists in many fields have come under attack for neglecting to consider the wider social and moral implications of their work. Yet those scientists who saw the light and decided to save the human race, despite itself, have created problems of their own (Weindling, 1989), as the history of nineteenth- and twentieth-century demographic medicine and eugenics demonstrates. Fabian politicians in Britain founded research and educational institutions devoted to this theory. Comte, the nineteenth-century philosopher, believed that government could be carried out by a committee of wise men, trained intellects removed from the interests and errors of politicians. Despite the very real achievements in some fields (and at least two of the early ecologists I discussed in *Ecology in the Twentieth Century* were Nobel Prize-winners in physics and chemistry) of scientists who have turned their attention to managing the planet's affairs, scientific achievement in one field is demonstrably not a warranty of clear thinking or right judgement in another. Several writers on deep ecology give their work a scientific gloss, using terms drawn from theoretical biology or cybernetics, and illustrating them with graphs and models. The certificate of scientific origin is crucial to the effectiveness of the ecological argument.

One response to this analysis may be to argue that Green thinking is not necessarily accurate at all, or clear, or scientific, but is unthinkingly apocalyptic, exaggerates figures and uses them selectively. In short, Green argument does not conform to the rationalist model of scientific thinking. True – and I think that without giving examples which would only cause unnecessary pain, one can point to Doomsday prophecies of the 1960s and 1970s which have not been fulfilled, despite being issued by respectable professors at universities or research institutions (McCormick, 1989). Nevertheless, Green argument does depend to a surprising degree on 'scientific' facts. It depends on the economist's calculations and the climatologist's assertions. Although New Age ecologists do exist who genuinely support a primitive and ascientific new religion, they are rare.

But what about the consideration that there are *real* environmental problems, that pollution and other catastrophes loom over us so horrifyingly that the methodology by which we become aware of them is irrelevant? The ecologists' argument is that even if these disasters lie in the realm of speculation, the situation is so threatening that action should be taken to avert them. There is no space in this book to discuss the likelihood or otherwise of the multitude of threatened ecological disasters, but it is not necessary to do so in order to suggest convincingly that ecologism and real disasters have little connection. Single issues, the concern of reformist environmentalists rather than political ecologists, are dealt with anyway, in the normal course of political lobbying. They will probably be dealt with piecemeal, and probably not globally; but such a procedure at least defuses the possibility that drastic global steps will be taken that may turn out in later years to be absolutely wrong. The ecological package-deal is not a response to real problems at all.

To explain what I mean, take the greenhouse effect. One might say that the ozone layer put Greens on the map. Since other disasters have not yet materialised (there is still standing room on the planet though population levels are still increasing, we are still discovering oil reserves), the discovery of holes in the ozone layer came just in time, and the greenhouse effect, while still in the realms of prediction, has galvanised governments into action.

But that is the point: it *has* galvanised governments. If there had never been an ecological movement, and evidence had emerged to suggest that the greenhouse effect existed, and that some countries might be deleteriously affected by a greenhouse effect, then those communities who thought that they would be affected by it in a bad way would be lobbying to have something done about it. They might have problems in getting an instant response and instant action from the rest of the world. The burghers of Holland, facing a higher sea level, might well be attacked as selfish by those who were delighted at the prospect of a warm, dry Britain, with vineyards growing in Yorkshire; while those who prefer oaks to conifers could return to a Sweden their ancestors left over a thousand years ago. Nonetheless, there would be protests and there would be action. The protests would not be tinged with righteous hysteria, as now, and we would be much clearer as to what exactly was happening, or predicted to happen. We would not be told that this change of climate was a unique event since man first walked the earth, nor would distorted and exaggerated figures be bandied about as much as they are now. Nor would the imminent disaster fit nicely into an ecological package-deal, and therefore reinforce it, because such a package-deal would not exist.

Nonetheless, politics would have to be very inflexible and unresponsive, the future sufferers very passive, for the issue not to be taken up at all. For ecologists to argue that a possible greenhouse catastrophe is a vindication of their cause is therefore unreasonable. This argument holds good

whether or not one points to the numerous and contradictory climatic disasters that have been forecast for the past
thirty years. (The greenhouse effect still remains in the
realms of hypothesis, but one should accept on methodological grounds that human behaviour *could* cause a
climatic catastrophe, and any discussion of ecological
apocalypse should be prepared to acknowledge that fact.)

To explain further why I argue that the ecological movement does not depend on 'real' problems, here are two
examples which show that ecologism can arise regardless of
such events. As I have shown earlier, in the nineteenth
century, well over a hundred years ago, ecological arguments similar to today's were put forward. An anti-
Western ethic identified civilisation with exterminatory
and evil impulses, preferred matriarchy over patriarchy,
preached anti-anthropocentrism, thought an experience
of the wilderness essential to a viable society, talked in
organicist metaphors, and above all believed that resources
were finite, and must be shared and rationed (Bramwell,
1989). That sense of doom, of apocalypse and of guilt,
which is so strong in today's political ecologism, was just as
strong then.

But it would be hard to argue that real environmental
problems sparked off this ideology, or that there was any
sign of resources running out. Were these early ecologists
cleverly predicting what has happened now? No, because
the disasters they sensed have not happened yet. We have
not yet run out of land, space, water or energy. Naturally,
all resources are finite, leaving aside for the sake of the
argument man's technological or inventive capacities. But
thinkers who harp upon a possible future danger to our use
of resources, but offer only an indefinite future chronology
and an indefinite causation, are not being very helpful. Nor
am I picking on particularly second-rate or eccentric
thinkers in looking at these early ecologists. Establishment
people are as prone to unreality, to the dictates of fashion
or politics as anyone else. To rise in the establishment you
do not take an exam that guarantees you against ever

having any dotty ideas; you simply take an exam that ensures you have the same dotty ideas as your peers. Nineteenth-century ecologists shared an intellectual world, for the first time, and were thus especially vulnerable to sharing the political passion that arose from their fears and loves.

The second example is that of Eastern Europe today, which I will discuss at greater length in Part III. Everyone knows by now that pollution in the Eastern bloc and the former Soviet Union is appalling, and they probably know that environmental protest was a major factor in pulling down the ruling communist regimes. Yet the ethic of political ecologism can be found today in one East European country only, and that country is Bulgaria, where the rhetoric of technophobia, talk of cultural domination and a return to spiritual roots resembles the deep ecologism of Western Europe. Bulgaria, however, is the East European country with the *fewest* environmental problems, almost the lowest levels of sulphur dioxide and nitrogen dioxide pollution, and the lowest level of industrialisation. In seriously affected countries such as Czechoslovakia and Poland the urgency is real, and the answers pragmatic, a practical approach to environmental problems that cannot be described as ecological or Green thinking. So we are left with the paradox that the self-assurance of deep ecologists, their moral justification, the certainty of their cause that impels them towards direct and violent action, is based on a belief in the scientific validation of their position, a certainty that true and right reason, a dash of holism, a soupçon of vision, will turn bad science, positivist and analytical, into good. And no doubt scientists in their turn find in ecological causes a way to fill the void created by the rigours and boredom of their discipline.

Just as the attitude of deep ecologists to science is contradictory, so their approach to the state is ambiguous. To make civil disobedience effective (that is, to persuade a government to carry out the policies you desire), certain preconditions are necessary. These include, obviously, a loss of faith in normal political processes, and a belief that processes are irrelevant compared to ends; but more importantly, proponents of civil disobedience assume that dialogue with the state is possible, that certain moral assumptions about fair play and sincerity will prevail. Civil disobedience assumes that the state will not use the power which it has available. It is in a sense a challenge to the state to do so, but if the state accepts the challenge, then its superior force will ensure victory, and sometimes the state does do so. Gandhi's tactics worked with the British in India, but failed when the Indians took over their own government. Hunger-strikers were pardoned by Tsarist Russia, but murdered after 1917. Nonetheless, the ecologists' belief in effective civil disobedience implies a certain trust in the state, trust and resentment, an attitude that parallels that of youth to its parents. It implies a position of powerlessness, of immaturity, and the concomitant certainty that the youth will eventually be able to take over from the discarded father.

Deep ecologists have given birth to the ecological saboteur, yet the logic of their meditative and introverted ideology seems to be against violence. The man who coined the concept 'deep ecology' is a Norwegian philosopher, Arne Naess, and his writings are crucial. They emphasise the role of action, and have influenced a generation of young American ecologists (Naess, 1989). Later writers on deep ecology and ecological ethics have avoided the question of conflict. They presume the possibility of a conflict-free world, based on a benevolent nature. Bill Devall has written that deep ecologists, unlike reformers, think of the inclusive well-being of all living beings (Devall, 1990).

Plenty of ecologists who do not accept the whole 'deep' package do instinctively accept the Aldo Leopold principle, that you should live in such a way as to respect and maintain the integrity of the existing eco-unit. This means respecting *all* its components – at least, Leopold does not give us any idea how conflicting claims should be resolved

(Leopold, 1968, pp. 225–6). Ecologists do not always see why there should be conflicting claims at all, assuming that the answer is to confine the effects of human intrusion to some clearly establishable and obvious minimum, while refusing to specify what that minimum might be.

The picture here is muddied, because of course man is a part of the ecosystem in any case, and was put there by mother nature, not society. But the essential point is that an ecosystem is not an optimum and unchanging unit, as the proponents of the 'climax' theory believed. It alters: the balance of species within it changes, and it does so without human intervention. Furthermore, an ecosystem runs on a constant transfer of energy from species to species, which is performed by consumption. Animals eat animals. Animals eat grass. The food chain means the utilisation of one species by another for survival. What holds back the growth of one species over another is shortage of food. That is what the balance of nature means. When we interfere with this delicate balance, the problem is to determine at what stage we stop, and how we weigh the relative interests of the species we interfere with, not just interests relative to ourselves, but those relative to other animal species.

The doctrine of animal rights was made famous by philosopher Peter Singer, who declared that he was unaffected by the charms of the animals concerned (Singer, 1976). Spiders had rights if dogs did. This rigour soon became muted among Singer's followers, because apart from its impracticality the desire to give animals rights is regrettably tied up with the respect and affection we have for some animals, and mammals tend to receive more of this human value than do fish, insects and bacteria. The deep ecologist belief in full biological equality has been tempered by the appearance of a subtly argued programme for *mammalian* rights (Regan, 1988).

Arne Naess does discuss these questions. To give one example, he asks if a farmer is justified in killing an animal that worries or kills part of his flock. Unfortunately, his answer avoids the problem. Naess argues that a careful study of such situations shows that the conflict is only apparent. In real life, the villainous animal will often turn out to be a rogue, or to be behaving abnormally, or to be the victim of the farmer's mismanagement, and hence not responsible for its actions. In situations where the claims of pastoral life conflict with the claims of the indigenous animal life, the latter should come first. But, adds Naess, the issue is a complex one, and needs a great deal of study. There are shades here of 'Needs Further Research', that stock finish to a paper funded by the Economic and Social Research Council. After so many years of *thinking*, one is surely entitled to a clearer conclusion about conflicting man–animal claims, or quantified suggestions about exactly how much we can use of any resource; otherwise, why should one trust the Green planners? And when it comes to the claim that wildlife should be respected more than domesticated life, we are entitled, I think, to something more specific than a holding operation while we look for the sparking plug.

One reason for the failure to have finalised the deep ecological idea is that the very values which make people want to respect the rights of mammals and trees operate against real egalitarianism. Love, precisely because it arises from differentiation, and also from the differentiation between subject and object, *is* hierarchical. So elephants and pandas seem to us to have more rights than spiders, and spiders more rights than bacteria. The belief in stewardship that motivates many ecologists is paternalist, and suggests that human values and human survival have a dominant role to play, even if such values should be changed to include love and respect for nature.

But a deeper reason lies behind this failure to produce concrete recommendations, despite decades of intensive thinking. Andrew Dobson has written of the strange reluctance of ecologists to communicate solutions, especially in terms of means (Dobson, 1990, pp. 22–3). What they like to do is to indicate that the problems are being considered, even if solutions have not yet been found. It is as if they prefer an open-ended process of consideration to a final set of principles. The charitable interpretation is that this is due to their democratic and discursive habit of mind. Ideas are offered for consideration by co-operative and like-minded peers. This is common in party politics: you do not offer your manifesto for criticism by the opposing party. But ecologists are not a party. The introverted nature of ecological discussion involves only the converted. Clearly, this process is seen as important to deep ecologists. The language used in describing the evolution of a final set of principles is significant in this context. In Bill Devall's book, the *process* of evolving answers is referred to a great deal. On almost every page he talks about it: deep ecologists meditate, evolve strategies, study conflicts, invent terms, offer initiatives, provide guidelines and originate principles.

But another reason for the need for open-ended discussion is that some of the suggested ways forward are fatally flawed. It is all very well for the utopian socialist to talk about decentralisation, and/or democracy at local level; past experience shows this can be a code for rule by local mobsters, or we can take it as rule by a kind of glorious and continuous street party. In any case, it is outside the realm of reality, because Green programmes demand a powerful central authority to enforce them (Dobson, 1990, pp. 127–9). One suspects that the 'people' invoked by deep ecologists are the people who run state-aided opera houses in Nicaragua, not the 'people' who go short of grain because the train-drivers are on strike. The means of the deep ecologist suffer from the same kind of vagueness, because the murky details of costing the basic minimum wage, or of making public transport preferable as an option to private, are beneath his interest. He is not that sort of bloke. The ends are clear enough: a clean, purified earth, a controlled and minimised human population. But in deciding how to attain them the ecologist prefers to evolve strategies and ardently to put forth proposals, because the dilemmas are so great as to be in effect insoluble.

The belief in education, decentralisation and a dialogue with a state that basically believes in the right approach is part of the amiable liberalism of the backwoods academia where much of deep ecology is located (and much of its libertarian wing). That is not a criticism either of their ideas or of their location: political ecology is a new and evolving discipline, and so is unlikely to be placed in establishment institutions close to the throbbing heart of urban life. But it might help to account for the vagueness about procedures. The idea of local democracy may seem to mean more in a true locality, in Seattle or in Toronto.

In the course of explaining how his ideal society is to evolve, the populariser of deep ecology, Bill Devall, tells us to 'think like a mountain' and 'sing like a river', become 'one with the earth' (Devall, 1990, p. 191ff.). That way, you can avoid the painful conflicts and choices, because man is cruel, but never nature. But how is this conflict-free oneness arrived at? Suppose that by meditating you acquire the characteristics of mountains, rivers and the earth. Why should we suppose that these characteristics are bound to be benevolent? The concept is not known to natural phenomena: it is a human characteristic, and assumes the power of choice, of good but also of evil. The belief that you can rise above cruelty and kindness is actually rather troubling. For what lies behind this image of the human being above paternal benevolence, the man partaking of nature? Not only is nature not benevolent, but in the long term neither are her effects. Volcanoes violently alter local climates. Comets have the potential for wiping out all life on earth. Oxygen-breathing life exists as the result of the extermination of our non-oxygen-breathing predecessors by some unimaginable catastrophe. The uncaring river flows: it knows not time nor change, it is and it is not. Can humanity become like this? Should humanity become like this? It would be a very different species from the one we know now, and the obvious Green response – 'and a good thing too' – begs the question: would a humanity as supra-moral as 'nature' be desirable?

Devall and others would deny that human beings give nature its meaning from outside. But the argument for a non-human-centred world which has yet been most extensively developed, the Gaia hypothesis, which gives the earth its own vast lifespan, its own survival instincts, does not suggest that humanity should rely on nature.

One quality stands out in this image of the wise guru meditating on the mountain, in superhuman control of himself, the guru so admired by deep ecologists in the USA because of his contact with the higher powers, his *capacities*. He has created new paths to domination – tapped forces, but of course on an acceptably non-material level. This man is master of the universe. Indeed, Devall goes on to talk of the deep ecologist as warrior, and lovingly describes the techniques of aikido: 'A warrior trained in aikido advances and controls the center' (Devall, 1990, p. 191). Although Devall describes his eco-warrior as having tender and intuitive qualities, the image is that of the superman, in touch with the currents of the universe and able to use them as functioning strategies. Devall goes on: 'By becoming empty, fully empty, Buddhist teachers say, we become full of possibilities.' *Possibilities*? To do what? Am I alone in finding something unattractively instrumental about this language?

Not all deep ecologists have followed this path. Arne Naess, a founding father of deep ecology, does not suggest tuning in to nature's powers in order to find the strength to see off the logging industry. The aim of his philosophy, which he calls 'ecosophy' (from *sophia*, the Greek for wisdom), is to demonstrate the interconnectedness of man and nature and to provide principles for ecological living. He uses the word *Gestalt* a lot, and anyone who agrees with Allan Bloom that taxi-drivers should not talk about *Gestalt* may well feel, by the time they have finished Naess's works, that neither should philosophers (Bloom, 1987). To Naess, conflicts become merged in the ideal of a holistic universe. No doubt, conflicting rights and claims do appear insignificant from a cosmic perspective, but that does not mean they do not exist, or should not be considered. Still, Naess's calls for rationality, common sense and restraint are attractive. He tends towards non-action, because human attempts to cure environmental problems so often interfere with the delicate ecological balance. The right way is to acquire an intuitive understanding of nature by living in the woods and fields and studying them directly, at close quarters. Naess is an existentialist. Learning by doing: action not words. He attacks the mainstream of European philosophy – Descartes, Bacon – for the same reasons ecologists attack what they claim is the mainstream of European science, arguing that it is mechanistic, analytical and reductive. This representation of European philosophy ignores the existence of Spinoza, Kierkegaard, Goethe and Bergson, but appears constantly in the works of ecologists, who see themselves as rescuing humanity from error. A certain ahistoricity is essential for this vision to survive.

Ecologists believe that orthodox science does not allow for values, is purely positivist and ignores problems of relativity. Yet paradoxically, those who call for a new holism are drawn almost entirely from the scientific community. And Naess, the deep ecologist, stresses the value of reason, because he believes that if his views are calmly argued, methodical and backed by fact, they must and should convince. So scientific backing for ecological ideas is a crucial justification.

It is not obvious why measuring the atmosphere's components and tracking changes in climate should be considered holistic science, rather than analytical, nor why counting species, alive, dead or endangered, should be a virtuous activity when other kinds of science are evil. Ecologists have not resolved this conflict – that science is good when it claims to discover a greenhouse effect, and bad when it invents chemical fertiliser.

In two other areas the world-open, value-laden philosophy of deep ecology seems worryingly fuzzy. Take one example. Naess, like most ecologists, argues for stasis, and against uncontrolled change. Above all, mobility is an agent for instability, and should be minimised. Man must be rooted to the earth, and the capitalist idea of worker mobility opposes this ideal.

This deep dislike of dynamism, uncertainty and change is understandable (even though it conflicts with the Green demand for free movement of peoples). There is a link between people and place, locality and continuity. But how is one to turn this into practical politics? Ban labour mobility? Discourage it? Naess suggests that one answer is to make 'the local community the natural starting point for political deliberation'. But how does that help?

First, there has to be a homogeneous community, agreed on its basic values and aims, which will tie its participants by subtle and non-coercive bonds. If such a community is not already in existence, then what? Naess says that the man-to-place link is easier to sustain before 'a policy of sustained material growth' has begun. This suggests that the correct ecological solution can only work in a very few areas of the earth, among residual and passive tribes. Presumably Naess did not mean this. He is, after all, talking of a generally usable model of community organisation.

So given your viable local community, it decides democratically how it is to survive, without sustained material growth, without losing members or gaining them. But suppose the community decides otherwise? Ecologists do not address this problem. The prospect of ineradicable disagreement does not enter their model of community.

This may seem surprising. After all, ecologists say that we are in the state we are in because certain human drives have got out of hand. Why do they believe that such disruptive patterns will not emerge on the micro-level?

One answer is their deep-seated belief in the appeal of community – *Gesellschaft*. Once offered such charms, who would abandon them? Unfortunately ecological philosophers are a self-selected group, frequently emanating from the university intelligentsia, though sometimes dropouts from it, and linked in a global village of believers. Since the hostile disbeliever seldom enters these villages, the philosophers are able to ignore their presence, and carry on believing that education in their values will overcome the rogue element in humanity.

But in any case, in pursuit of stability, something else important is written out of the ecologists' model: humanity's wandering impulse. Naess veils this. He attributes mobility, the drive to explore, to roam, to move, to leave, to a purely exogenous policy of material growth. Yet we have been peasants for a very short time indeed, set against the timescale of the human species. Nomadism pre-dates the settled world of agriculture by hundreds of thousands of years; and even the rooted peasant has proved surprisingly apt to wander about, even if only to the nearest stretch of vacant land.

So how would putting the roots–mobility conflict into the hands of the local community help matters? Naess opposes having a powerful local bureaucracy. But how else would an 'ecologically responsible policy' function? Ecologists talk of the need to put community before profit. But who would choose which community would be subsidised, and by whom? How would cross-subsidies work in a decentralised system? Communities can be geared to environmentally damaging functions. This dilemma has underlain orthodox paternalist policies in Britain, which have traditionally called for the maintenance of communities formed to extract coal, or work in iron or steel – polluting industries gradually discarded by market forces.

When it comes to restricting personal mobility to what is necessary, if market mechanisms are not to be used (road pricing, fuel taxes) then how would the model work?

Would someone who wanted to move from the Highland eco-zone to the Chilterns bio-region have to ask another's permission? Who would police the comings and goings of these local communities, and control the policies? These queries may seem crude set against the desire to 'get the qualities back into nature', to restore faith in 'our rich world of the senses' that Naess attractively propounds, but they are relevant to it. The conclusion to me seems inescapable: deep ecologists are holiday ecologists. They are vacationing in nature, and yearning for community, before returning to their protected lives as children of the state, a state all-powerful, but badgerable like a busy father reading the paper, a state which finally, like an aged parent, becomes disposable.

This interpretation is strengthened by my second example, of an ambiguous approach and geographically skewed blindness to environmental problems in the former communist states. For Naess entirely ignores ecological problems in the then Soviet Union. Naess did attend conferences in the Soviet Union, and should have observed the situation there, and might even have searched for information from protesters on the spot. He claims that 'at least in the Soviet Union the problems ecological movements have to face are not so different from our own'. He can hardly mean that those who write articles about pollution in the West are imprisoned for decades, or put into lunatic asylums, the fate of most environmental protesters in the Soviet Union between the brief thaw of the 1960s and the thaw of the late 1980s? He may mean that technology is the villain, and exists in all advanced industrial nations, whatever the system. He may mean that the environmental problems are not so different. But it is a problem that deserves closer examination by him, because it indicates that capitalism, the market economy and multinational corporations are not solely responsible for environmental problems (Seabrook, 1990). Other works on global environmental problems have also failed to address the problems of the former communist bloc (Paehlke, 1989). When a global movement with an ethic based on Planet Earth omits about a third of Planet Earth from its calculations, and by a strange chance that part suffering worst of all from the problems that are supposed to back up the ecological ethic, one may be permitted, despite the metaphor's 'speciesist' overtones, to smell a rat. Are ecologists really attacking pollution and overuse of resources? Or are they targeting the Western world and the market economy *per se*?

Naess may seem on firmer ground in his belief that reform should be accompanied by 'a value-laden appreciation of the human condition'. But his programme of non-violent civil disobedience relies on having a well-meaning and civilised opponent, a state which is fundamentally on your side. Civil disobedience works only when both sides accept the rules, and the moral principles behind the cause (civil disobedience would not have received much consideration in Britain during the Second World War).

These criticisms may seem harsh, directed towards deep ecologists who, after all, are arguing for spiritual regeneration, in the cause of the planet. We all want to save the planet. That is, few, faced by an opinion poll, would declare themselves to be anti-planet, or pro-planetary destruction, just as few would declare themselves as being in favour of nuclear war. Nonetheless, we do not all support unilateral disarmament. And those of us who wish to save the planet may legitimately disagree about the aims, methods and consistency of the self-styled protectors of the planet.

Since much of the deep ecological argument has somehow crept into the public awareness it clearly meets a psychological need: a need for values, for closeness to nature, for guilt and responsibility, for a meaning to life; in short, for what religion used to offer. Certainly, it would be a pity if the causes of environmental improvement and conservation were to be neglected because of the extreme claims of ideologues. Perhaps it is true that reform environmentalism would never be sufficient to force us to put natural beauty before our own needs, although it should be sufficient to prevent man-made disasters. Yet when ana- lysed, the deep ecological claim to moral truth seems thin, the comprehensive world-view tattier at close quarters than we are entitled to expect.

Ecologists should try to live as they preach. Deep ecolo- gists certainly believe this. But although it is certainly hard to be a palaeolithic hunter-gatherer all by yourself, there should be some kind of link between saying and doing. Do ecologists absolutely have to travel everywhere by aero- plane? No, they could go by boat. They could walk, not drive. But they do not. This is not just a cheap crack at Gucci Greens. The point is that the reason they do not do as they say is important: it is because they feel their time and energy has a cost, that it is a value which they are prepared to exchange against their other values. Technology, like capitalism, involves a trade-off, a process of evaluation, as well as of monetarisation. Perhaps deep ecologists should meditate on this point.

CHAPTER NINE

Ecocentrism: A Viable Ideal?

ONE OF THE key elements of ecologism is its opposition to anthropocentrism. The alternative that has been evolved is 'ecocentrism'. As discussed in the section on deep ecology, this is a concept that has run into difficulties. What are they? One of the most respected historians of ecocentrism is Roderick Nash, author of *The Rights of Nature* (1989). He argues that the idea of 'rights' is one that has expanded over the centuries. Just as people used to deny rights to blacks, or to women, so rights have been denied to nature, and just as rights have gradually been extended to blacks and then to women, so they are now being extended to nature.

The concept of rights here is a difficult one. The kinds of rights extended to blacks were legal ones, and rights extended to women were fundamentally legal ones, too. Blacks gained rights over their own labour and persons; women gained property and voting rights hitherto denied them. These were highly specific accessions to the individual's civic and legal persona.

It is hard to see how rights of this kind can be applied to inanimate objects. Rights carry the obverse burden of responsibility. Legal rights accompany legal duties. One might argue that law can apply to inanimate things or legal fictions. For example, legal personality can be attributed to non-individuals, such as companies, and these legal fictions can then be held liable in civil law. Also, individuals who are incapable of representing their own interests – young children, lunatics and so on – can have legal standing, and have rights which can be protected in law. But to operate a reductionist critique of the individualist basis of law does not take us very far when it comes to defining the rights of nature. Nash does not deal with the problems of the idea of nature's rights. His is a history of the individuals who have believed in it, with a critique of those who do not do so.

One of the most fundamental rights we recognise is the right to live. This is not the same as the right to existence, which is the right involved in the case of inanimate objects. We do not say that a building has a right to live. Nor do we say that a building has a right to exist. However, though inanimate, a building is artificial, not natural. Let us consider the rights of a rock. We need not necessarily deny a natural object the right of existence, although the definitions and values have to be attributed by us. But other problems arise. If a rock is alive, then the very nebulousness of the concept 'alive' as aplied to a rock makes it hard to prove that you have killed it. Even strip mining will merely rearrange the atoms of a rock. Non-existence for a rock is different from non-existence for a conscious or otherwise animate being.

If we accept the right to life as a fundamental right of nature, then if it is to mean anything other than a rhetorical gesture it must mean that we do not destroy natural artefacts. This has been interpreted to mean the preservation of rivers, for example, from the building of dams or from pollution, and of trees from logging. But if we are not to destroy rivers or trees, then presumably we should also not destroy animals and non-tree vegetation. If a river has rights, then a lettuce has rights. Similarly, Environmental Law may propose a constitutional clause to the effect that 'wildlife must not be deprived of "life, liberty or habitat without due process of law"'(Nash, 1989, p. 6). But why should this protection be confined to *wildlife*, and not extend to tame life? From this desire to protect the domesticated natural world comes the idea of permaculture. By

cropping the tree, you do not kill it. Apart from the energy
savings involved in permaculture, the moral aspect is important to ecologists. It still involves stealing the produce of
the tree, the cow, the brussel sprout plant. Permaculture
may be an answer to the conundrum of respecting the
rights of vegetation. However, I do not think it would be
possible to sustain the human population of the planet
using this principle to produce food. Indeed, the alternative
to anthropocentric ethics (mischievously compared by
Nash to slave-owning greed) is total planet death. Nothing
would be allowed to eat or consume anything else. We do
not permit cannibalism. This is first among the rights we
afford ourselves: that we shall not be eaten by others.

Nash comments that criticisms like this are unreasonable.
After all, 'human-to-human ethics have not been entirely
clarified' (ibid., p. 7). Incredulity, he notes, met 'the first
proposals for granting independence to American colonists,
freeing the slaves, respecting Indian rights', and so on.
However, the degree of incredulity aroused by a proposal
does not guarantee its soundness, or its eventual acceptability. Flat-earthers meet today with incredulity. We have
to look for something else to convince us.

If we give a tree a legal right to live, it is a measure that
operates through the application of human values. We are
not recognising its right, but recognising our moral duty
to protect and preserve. But ecocentrists believe that
'nature's' right to survive is absolute, and embedded in
itself. Humanity must take its place as an equal but not as
a superior in the ecosystem. If conflict arises between human needs and those of the ecosystem, then the ecosystem
should take precedence.

Ecocentrics are among the most consistent and least
democratic of ecological theorists. They argue that if
democracy cannot solve pollution problems and prevent
the greenhouse effect, then 'new forms of government'
should develop that can do so (ibid.). The word 'new' is a
formulation often used to obscure meaning or veil a lack of

meaning, and its frequent use is a characteristic the Greens share with the pre-war European Fascists. Like Arne Naess's solution to human–animal conflict, and the formulation used by Niklas Luhmann in his *Ecological Communication*, it can mean everything or nothing.

That is not to say that the idea of allotting rights to nature is a vacuous one. But the true comparison is not with the liberal theorists, the Benthams and John Stuart Mills, but with religious concepts. A comparison I proposed in an earlier book (Bramwell, 1989) was between ecologism and the ideal of chastity. The ideal of the non-sexual life arose from the same self-abnegatory and guilty roots as anti-anthropocentric ecologism. It was an impossible, confused, self-defeating ideal, but a powerful one. To this day, some men and women live in separate communities and take the vow of chastity. With ecocentrism, instead of vowing one's libido to God, we vow the libido to nature.

PART THREE

Green Futures

CHAPTER TEN

Environmentalism in Eastern Europe

THE GROWTH OF an environmental movement in Eastern Europe provides a valuable case study for comparisons and contrasts with environmental movements in the West. Most of the sociological variables differ but there are two striking things in common. One is that the first manifestations of environmentalism in Eastern Europe flowed from political liberalisation. In the Soviet Union, for example, considerable discussion about pollution problems was permitted in the 1960s. It was noticeable that this was connected with a revival of cultural nationalism, an element that in Czechoslovakia was lacking in the brief period of reform known as the 'Prague Spring' of 1968 under Dubček. The liberalisation process was caused by an awareness of the moral and economic failure of the old system, whereas in the West it accompanied the awareness of failure of earlier left- and right-wing ideologies. At one level, this is not surprising. Given a relaxation on controls, problems could be brought into the open, criticisms made. It meant too that leeway existed for discussing social aims and priorities. All East European countries had constitutions with a clause stating that the exploitation of nature for the benefit of socialist man was essential. Thus, economic development had to take precedence over environmental protection, laws for which were gradually introduced from the mid-1950s on.

The second similarity is the involvement of the intelligentsia. Just as political dissent in general was headed by the educated, so this particular topic was dominated by the university educated. As in the West, scientists became concerned about pollution problems. However, these problems were not the apocalyptic visions of the future that possessed Greens in the West, the fear of running out of finite resources, the desire to live more simply. They were problems that could be seen around them. Public health was visibly endangered. Death rates in some areas were increasing. Children were suffering.

Furthermore, unlike the situation in the West, dissent over environmental problems arose as the economic situation was worsening. It was therefore not a response to economic satiety. It was closely connected with the appearance of real problems, and had a hard-headed quality that was less apparent in the West.

Environmental problems in many countries were energy linked. The decision to go for economic self-sufficiency during the 1970s had meant that deposits of high-sulphur brown coal (lignite) were exploited in order to minimise imports of oil, hard coal and natural gas. In Poland, hard coal deposits in some mines were close to being mined out; in others, mining the deep coal was getting increasingly expensive. Brown coal usually lay on the surface and could be strip-mined. Power stations were built next to the mining operations for cheapness. The result was an increase in gaseous and particle emissions. At the same time, the infrastructure inherited from pre-war times began to show its inadequacy. Water treatment plants scarcely existed. Industrial waste was discharged directly into rivers. Monitoring facilities were scarce and inadequate.

Dissident scientists and others concerned with the environment identified legal responsibility as one of the main problems. The state could not be sued. The state had a vested interest in ignoring problems created by state organs. Fining state-owned factories was simply a transfer payment. So the introduction of private property became a prerequisite for improvement. Law journals in Hungary followed new ideas for economic instruments in the West with interest, publishing eager articles about the latest 'polluter pays' idea. Where Western Greens were prone to attribute environmental problems to greed and materialism, the East European environmentalists attributed them to aspects of the socialist state: centralised planning, the Soviet domination that led to unnecessary development of heavy industry, the failure to price resources and the difficulty of costing them, the aim of high production targets regardless of environmental impact, the low priority given to pollution abatement, and the lack of individual responsibility associated with the command economy.

The environmentalists in Eastern Europe identified another factor as responsible for their plight, namely participation in and subjection to Comecon, the trading system of the Eastern bloc, dominated and controlled by the Soviet Union. Under the barter system, countries swapped finished industrial goods for raw materials from Russia. They had an economic incentive to over-produce shoddy goods which could be dumped in the Soviet Union, a process which led to considerable waste and considerable demoralisation among the workforce. Under a Comecon system of division of labour, countries had to produce so much steel, iron or grain. One consequence of this was that heavy industry was developed in unsuitable areas. Large steelworks were built next to the historic city of Kraków. Czech agriculture was turned over to mass grain production. Some dissenters believed that the degradation of historic towns and landscapes was a deliberate and political act. Kraków, for example, had been part of Austro-Hungarian 'Congress' Poland before 1917, a prosperous, bourgeois area. It was felt that the vast steelworks of Nova Huta had been erected expressly to annoy the remnants of the Polish middle classes, and to deface a historic part of Poland's heritage.

Similar feelings were expressed in Slovakia. Industrialisation and road-building were concentrated in the east of

Czechoslovakia. Possibly the intention was to move the economic and industrial centre of the country further east, towards the Soviet Union, and away from German-speaking countries. But the process of industrialisation was resented by Slovakians, who perceived it as an attack on their landscape and way of life (the mountain areas of Slovakia resisted collectivisation much longer than the rest of the country).

The Eastern bloc countries thus could not control their own industrial production, or choose a mix of goods that would minimise pollution problems. What they could choose to do was to try to minimise side-effects by siting highly polluting plants on their own borders. The concentration of such plants in border areas in the former East Germany, Czechoslovakia, Poland and Romania is noticeable. During the Comecon era it was a source of constant, grumbling irritation, as was the transference of polluted waters from one country to another. Hungary, for example, is mostly flat, and forms a kind of sump into which rivers flow from Czechoslovakia and Romania. The main rivers flow north through Poland, and expel their waste into the Baltic. Before the great change of 1989 there were increasingly agitated meetings and conventions between the countries involved. There were protests in Poland about air pollution imported from East Germany and Czechoslovakia, while Romania and Bulgaria were similarly involved in quarrels over pollution emanating from the river that ran through their boundary, the Ruse.

Environmentalism, therefore, became the focus of a general political syndrome of dissent, which included many members of the ruling system, the *apparatchiki*. The lack of economic and political independence of the Eastern bloc; their attachment to a tyrannical and more primitive overlord; the failure of the system to respond to individual needs, whether as consumers, workers or citizens; the endemic secrecy of the system, its backwardness compared with the West, and its hostility to historical and cultural values: all helped to turn the landscape of much of Eastern Europe into an environmental disaster area.

So environmentalism in Eastern Europe did not spring from some nebulous and unfocused sense of dissatisfaction with the modern, industrialised world *per se*, but rather from opposition to, and concern about a quite specific form of forced industrialisation, and its all too visible and tangible destructive consequences. Yet a process more similar in some ways to that broadly evident in the West can be seen in Romania and Bulgaria. The two least industrialised countries of the Eastern bloc were confronted with forms of protest that were more impassionedly cultural and nationalistic than in Poland, Czechoslovakia or Hungary, protests that focused much more on the virtues of the peasant way of life. One of the first expressions of dissent to reach the West from Romania, smuggled out in 1985, came from a dissident, Doinea Corneu (later to resign from the National Front after Ceausescu's fall) who spoke in an unpublished declaration of the destruction of villages under Ceausescu, and wrote movingly of the importance of the peasant to the nation, which she described as rooted in the peasant's religion, folkways, buildings and care for the land.

In Bulgaria, the least industrialised nation of all in the Eastern bloc, but one of the most fervently nationalistic (Bulgaria *irredentia* still having designs on Macedonia, some of Romania, etc.), environmentalism has a fundamentalist tinge much closer to the deep ecologists of the West. Bulgarian environmentalists oppose industry and capitalism.

The amazing revolutionary events in Eastern Europe in 1989 have passed into history, and attention is now focused on the diverse processes of rehabilitation and reconstruction, the painful adjustments to a market economy, the attempt to create private property and settle old claims. Successive socialist governments fell before any coherent idea had developed about what political forms might replace them. The collapse of the regimes left an inheritance of infrastructural decay and poverty that shocked even the protesters. Vaclav Havel, former dissident playwright and writer, elected President of Czechoslovakia in 1989, spoke in his New Year's Day message of the state of his country. What they had thought to be a 'disordered house' was in fact a ruin. What was to be the role of ecologists in this situation?

Protest against environmental degradation had played a major role in several of the protest movements in the Eastern bloc. For example, dissent over the proposed Danube dam in Hungary focused opposition to the government. The need for building the dam could be justified: Hungary's rainfall is low, both an increased water supply and the additional power generated by the dam would be welcomed. However, apart from the local damage to scenery and wildlife caused by the building of the dam, its effect on the lower Danube would be catastrophic, making the river more sluggish and dirty. The proposal led to opposition; the government accepted a petition against the dam, and held a referendum on the idea. When the referendum produced a verdict against the dam, the government decided to halt construction, even though this entailed compensation payments to Czechoslovakia and Austria, the other countries involved. This was the first time a communist government had responded to a popular protest in this way, and it signalled a readiness to negotiate and compromise that led to more widespread political change within months.

However, the ecological protest movement did not become part of the new political life. The dam activists split into different groups once the dam itself was cancelled. They found it hard to find a single issue that could mobilise as much support as the dam, did not have a general platform, and in any event did not think the time was ripe to form a party. Environmentalism in Hungary tended to be a thing of the educated classes, of the intellectuals. The fact that after the dam the next main issue was air pollution in Budapest (caused largely by the two-stroke-engined cars then in use), and the fact that the protesters were predominantly students and professors at Budapest's universities, exacerbated this image. When a Hungarian Green Party was eventually formed in 1990, there were suspicions that it included many former communists. In consequence, they did badly in the first free election in 1990, winning less than 1 per cent of the vote.

What then happened was a process similar to that which took place in several other East European countries. The diffuse but powerful urge to protect the environment was not translated into electoral success. The government did not include many environmentalists (Czechoslovakia was to do so, but their power was limited). In Hungary, economic growth was regarded as a prerequisite for environmental improvement. Powerful interest groups, such as the Hungarian mining conglomerates, were unresponsive to environmental demands. And as elsewhere the old bureaucracy was left largely in place. Although not particularly committed to communism, they were not responsive to new ideas either. In this case, the Hungarian Cabinet included the old 'water lobby', who had wanted to build the dam in the first place – the old Environment Ministry had been part of the Water Ministry, and setting up a separate Ministry for the Environment took several months. Finally, the crusading Green, Sandor Illes, who wanted an Environmental Protection Agency, was brought in by the Environment Minister, only to be subsequently dismissed for putting his environmental views ahead of those of the ministry. Illes had complained about the low priority given to restoring environmental damage in Hungary.

The anti-dam movement had been motivated partly by the historical importance of the Danube on the Austro-Czech border. Similar cultural and nationalist aspects to environmentalism in Eastern Europe had played a part in focusing dissent, as had distress at the neglect and destruction of the countryside.

CHAPTER ELEVEN

The Impact of Eastern Europe on Western Greens

IRONICALLY, SINCE THE fall of the communist regimes in Eastern Europe, the increasing awareness of environmental problems in Eastern Europe and knowledge of the role played in the revolutions by environmental issues, has somewhat destabilised Green politics in Western Europe.

In the few years up to 1989, there was increasing awareness of the extent of environmental degradation in Eastern Europe. Global monitoring of acid rain meant that polluting emissions from Eastern Europe could not be entirely hidden. Eastern bloc countries took part in conferences and signed agreements to reduce their sulphur dioxide pollution. Greenpeace and other pressure groups published data, guesstimates and anecdotes from dissident groups in the East which revealed an appalling state of affairs. Toxic waste was dumped and its location then lost. Water in rivers and wells was poisoned by untreated industrial, animal and human waste. Above all, the use of brown coal for the generation of power had led to widespread air pollution. This damaged health, particularly that of children. A number of reports of pollution and of hazardous accidents were smuggled out by dissidents.

The reaction of East European governments varied. For most of the time they tried to hide the facts, imprisoning or otherwise oppressing dissidents who wanted to publicise the situation. The Czech government commissioned studies of pollution in 1983 and again in 1989, but decided not to publish them. The Polish government from 1981 on admitted its environmental problems, but its Central Statistical Office produced data that downplayed the problem. Much of this pollution was exported to neighbouring countries. The Vistula poured out its dead waters into the Baltic. Romania's pollution affected the Black Sea delta. The river Werra, bordering East and West Germany, endangered West German soil and water. Research institutes in nearby countries affected by this pollution – West Germany, Austria and Sweden – tried to measure the extent of 'exported' air and water pollution. It became clear that East German power stations contributed significantly to the amount of acid rain in West Germany, while the Czech complex of mines and power stations in North Bohemia damaged forests in Bavaria. The political atmosphere was so taut that the Czechoslovak government refused the offer of a more advanced and clean power station, even though it was free.

Exact knowledge of the extent of the environmental problem was obviously hampered by government secrecy in the affected countries, and the media in the West did not find it a 'sexy' subject – it smacked too much of Red horror stories. However, visitors to Eastern Europe in the mid- and late 1980s included academics of all political persuasions, looking for amiable dissidents to grace their faculties. In Poland, especially, they found able specialists who had tried to keep records, albeit with inadequate monitoring equipment, but who could point to a worsening of all available indicators. Death rates from cancer and heart and lung disease appeared to have worsened since the 1960s. A stench, a brown smog, hung over the mining areas. Rivers even in rural areas were visibly dead. Historic buildings and town centres were damaged, some beyond repair. News of such catastrophic conditions began to spread, largely by word of mouth or through small-circulation journals.

During 1989, as one by one European communist governments fell or were weakened, pollution problems in the East became public knowledge. During 1990 the press in the USA and in Britain was full of reports of the extraordinary level of environmental deterioration of air, water and earth. In consequence, environmental improvement became a priority for the many aid programmes prepared by the US and the EC.

At one level, the opening up of environmental problems did not greatly change public perceptions of life under communism. Despoliation and inefficiency seemed to be par for the course in Eastern Europe, which was governed by uncaring, incompetent monoliths. The unsubtle masses were unsurprised. But there was another level at which the revelations had hit at important preconceptions about environmentalism.

As I pointed out earlier, Western ecologists had tended to ignore environmental problems in Eastern Europe. Activist pressure groups, most noticeably Greenpeace, had begun to publicise the export of toxic waste to Eastern Europe, and studies of acid rain contained United Nations monitoring data from Eastern Europe. However, the deeper the Green, the greater the tendency to turn a blind eye to the problem. Most ecologists saw the Western system as the root cause of environmental problems. The desire for growth inevitably meant greater use of the earth's resources. Greed, individualism and materialism were to blame. Capitalism was seen as especially exploitative, and late capitalism led to the centralised state and monopolistic multinationals. Everything had gone beyond the people's control. Give the world back to the people, and their good instincts would triumph over malevolent money-grubbing. It was hard to admit that the situation could be as bad or worse in states where individual property ownership was severely limited or abolished, and where capitalist forms of materialism did not flourish.

Apart from this structural barrier to recognising pollution under communism, the impulse to Green politics in Germany, the USA and Britain had, as discussed earlier, been a left-oriented one. After all, it was East European dissidents such as Bahro who had originally been received by many Western Greens, especially in Germany, as virtual saviours, free of Western guilt, and persecuted by the Eastern bloc regimes. Bahro had not come as a communist supporter, of course, but nonetheless could be described as a left-dissident, a utopian socialist rather than a believer in capitalism. He had rejected the inhuman monolith, the paternalist rigid state, in favour of a more sensitive 'soft' participatory anarcho-socialism.

Many West European ecologists envisaged a stratum of such intellectuals in the East, oppressed by censorship, seeking spiritual values, rejecting materialism, waiting to be rescued. They believed that the two systems could converge, and thereby be rid of their rigorous, materialistic aspects. The oppressed intellectuals in Eastern Europe were especially the subject of pity and admiration, as the bearers of culture and knowledge. West German Greens, in particular, with their strong pacifist tradition, clung to the belief that nobody in the East could want war, or could possibly support nuclear weapons. Germany, then still a divided country, was peculiarly sensitive to the danger of supporting revanchist sympathies, of seeming to whip up anti-Soviet hostility. After all, Germany was still the country where the *Historikerstreit* controversy could emerge, where historians taking an anti-Soviet stance were accused of perpetuating the Cold War, and even accused of trying to justify Hitler's invasion of the Soviet Union (Augstein, 1989; Evans, 1989).

The 'convergence theory' puzzled environmentalists in Eastern Europe, who turned to the bloc of Greens in the European Parliament for support. The French Greens, in particular, proved to be supportive, perhaps because of their traditional cultural links with Romania and Poland.

The fall of communist regimes in Eastern Europe thus wrongfooted many Greens in the West. In Germany, it meant difficult decisions over the future of the West German Greens: whether to absorb the East German environmentalists, or to leave them as a separate party. The diffused structure of the West German Greens made this less of a problem than it would have been in Britain, for example. In Berlin the Greens still stood as the Alternative List. With their different diagnoses and policies, it would have been hard for the two countries' Green groups to merge. In any case, in the first German unified elections of 2 December 1990 the West German Green Party's share of the vote dropped dramatically to 3.9 per cent, less than the amount needed to claim parliamentary seats, while the former East German 'Green' alliance won only 1.6 per cent of the vote. They were able to take seats in the Bundestag, but only because the electoral 5 per cent rule was abrogated for East Germany.

Unification effectively brought political life for the German Greens to a halt. First, it changed public perceptions of social, economic and political priorities. Secondly it filled the void of truncated identity that had helped provoke cultural criticism of the Greens. Thirdly, the moral and economic bankruptcy of the communist governments made the anti-capitalist, anti-Western stance of many Greens seem irrelevant and out of date. And while the German public wanted to open their arms to their brothers from the East, the idea of paying for it by tax increases and inflation was less acceptable. Their affluence once threatened, it appeared more desirable, more important than consuming fewer resources in order to save the planet.

Does this mean, then, that support for the Greens in Germany was never more than a displacement mechanism, a means of expressing a vague dissatisfaction, an expression of boredom and satiety? Not entirely, for that judgement would underrate the extent to which West Germany, a consensual society, has accepted environmental ideas and is putting them into practice in its legislation. If one were to analyse the appeal of West German Greens, three strands would emerge: a fundamentalist criticism (ecologism); a realistic criticism (environmentalism); and a left-liberal stance on social issues. The second of these has been taken up by the established political parties, especially at *Land* level. The 'Greenest' party in Germany today is the Christian Social Union. With a strong rural base, it represents a strand of old-fashioned love of nature and landscape that was not directly represented by the German Greens. The third strand has been temporarily submerged by the fall of communism. Contact with the more pro-capitalist and hard-headed Greens of the East meant that many Green supporters in West Germany lost some of their illusions.

However, the demands embodied in the first strand, fundamentalist ecologism, have not yet been met. Even within the Green Party itself the *Fundi–Realo* divide had given way to a new compromise position by 1990. It may be that the day of the *Fundi* has passed, that the caring young fathers and peace campaigners of today have moved on to new social manifestations of their beliefs: more focused, less draconian. But certainly the unification of Germany – and this means a new state, not a return to an old state form – has introduced enough flexibility, enough novelty to satisfy the most satiated citizen.

Has the emergence of an environmental movement in Eastern Europe weakened Western environmentalism rather than strengthening it? The answer is, to some extent, yes. In party political terms, the revelation that worse environmental problems existed under non-capitalist systems damaged the left-oriented Green parties, a consequence most evident in the results of the German elections of December 1990. On the other hand, in the USA, where no Green party exists, the effect was rather to legitimise environmental concerns in the eyes of the unsympathetic. Clearly, pollution in Eastern Europe could not be ignored as the invention of a lot of tiresome Greens. But if pollution was real, was a concrete phenomenon, in the East, how could it be ignored in the West? There might be differences of degree, and these were indeed relevant to the diagnosis, but the effect was to lodge the idea of a degraded environment more firmly in everyone's consciousness.

Certain groups within the environmental movement gained from these events. Pressure groups and non-governmental organisations who had begun to work with industry found they could offer expertise and contact with East European activists. Those, like Greenpeace, who had also been looking at clean process technology, and the Green Alliance, a British-based Green co-ordination group who had been working actively in Eastern Europe before 1989, were especially favourably placed. The balance of power among ecologists was thus tipped towards environmentalists and reformists, and especially towards those activists who were evolving specific programmes for industry. The perception that international action was necessary to help the situation also increased support for environmental legislation coming from the European Community.

Altogether, then, events in Eastern Europe strengthened the hands of environmentalist pressure groups and weakened national parties. They promoted the process of consensus and discussion between activists and the corporate sector that had been gathering strength since the mid-1980s. This is the quiet revolution that is replacing the anarchic utopias of the recent past.

CHAPTER TWELVE

The Green Future?

THE LAST FIFTEEN years saw the rise and the fall of the Green parties. They fell largely because the most successful Green party was the German one, which was untypically urban-left oriented, and locked into stereotyped reactions to the German problem. The brief success of the German Greens aroused unrealistic expectations in other national Green groups; unrealistic because the conditions that enabled the German Greens to acquire political status were unique to Germany and to German post-war political development. The inimitable nature of *Die Grünen* was not understood. Using them as a model hampered the natural development of Green parties in other European countries. They fell because events in Eastern Europe cast doubt upon some fundamental Green political tenets. Finally, they fell because the established political parties, together with international agencies, took on board those environmental programmes and criticisms that could be incorporated into established, institutionalised forms of political life.

It may be that national Green parties have outlived their usefulness. In Germany the Greens have abandoned 'Greenness' as a priority, and are set to replace the Free Democrats as a minority coalition group. Eventually, the *Realos* will oust or freeze out the *Fundis*, in order to play this role. In turn, reunification has changed the political map of Germany beyond recognition. In Britain, the determination to be democratic and indulge in the minutiae of party manipulation has prevented natural leaders from playing their obvious part. But the decline of party Greens, as remarkable as their equally rapid rise, is not just to be accounted for by the electoral and organisational problems single-issue groups face, or by the twists and turns of political life in each European country. Nothing has happened to change the importance of environmentalism in people's thinking – every opinion poll shows a steady concern for environmental protection.

It is, paradoxically, the very importance of environmentalism that has forced the main political parties to take account of it, and the adoption of environmentalist goals by international and supra-national bodies that seems to have rendered national political Green parties superfluous. As soon as major health and economic hazards could be identified, measured and quantified, the writing was on the wall for the secretive and uncaring approach to pollution that characterised governmental policy. Here, the reformist Greens, disliked by their more fundamentalist colleagues and criticised by Red entryists, could play a significant part. Industry is now beginning to adopt environmentalist policies, and it is the largest companies that are the most anxious to do so. The proposal hopefully put forward by 'Green capitalists' like John Elkington and Tom Burke (Elkington and Burke, 1987) that pollution control pays has been tested and found by many companies to be true, if only because environmental management is a powerful corporate tool for improving efficiency. If clean technology can eventually replace pollution control, then many of the demands of deep ecology, and the fundamentalist attack on growth, technology and industrialisation, will lose even more support.

The cultural criticism dimension of Green ideology is not likely to go away, but it is likely to be further marginalised, possibly dissolving into disparate occultist, matriarchal feminist and other similar groups. If so, the Western political system will once again have shown its flexibility, its capacity to neutralise and then absorb its opponents.

Greens will be left, as they started, with dissident scientists as their main driving force. But the contradictions of such a group mean that it will not support the development of political parties. The essence of politics is process and compromise. The degree of belief in their rightness that characterises Green thinkers will not allow for easy compromise. Furthermore, the values-based cultural criticism cannot be reduced to a programme – it is too individual, too personal and, in the end, too hierarchical.

The chief contradiction seems to be between Green exclusivity and Green egalitarianism. Both can be said to have Enlightenment roots and, like many Enlightenment ideas, they are pursued by Greens with medieval ferocity. Green exclusivity is a natural result of concern about scarcity: too many people and too few resources leads to the conclusion that people need to be rationed to match the resources. As soon as this fundamental vision is accepted, questions of How and Why enter the picture. An assessment of relative human value follows inevitably. The static nature of ideal Green societies, exemplified by ideas such as bio-regionalism, are relevant to this problem. A Green society is a self-contained society, working to keep change to a minimum, and attempting to control technological innovation and deployment, as well as to regulate market behaviour at every level.

There will be many Greens who disagree with this judgement and point to their egalitarianism, their belief in democratic participation and their preference for lowering resource use instead of population numbers. Yet many interpreters of Green philosophy have been worried by the paradox that apparently 'nice' Greens occasionally let slip ideas that do not fit the humanist ideals they otherwise espouse. Andrew Dobson, for example, in his discussion of Green ideology, sees ideas about restricting refugees and other incomers into the UK as a puzzling blip, a mode of interference on the bland screen of Green ideals (Dobson, 1990, pp. 96–7). But this blip will not go away. It is a

contradiction that will remain, a contradiction between
Green logic and Green emotions. While there are those
who see human exploitativeness as controllable, as a social
flaw, others who try to quantify the ideal ecological society
keep coming up against the conundrum that the anarchist
state is in some ways the most oppressive state, that the
freedom and anonymity that belong to the strong central
state are lost in the social pressure that keeps a tribal society
going.

The fact that every attempt to find out how a Green state
would work is met with a stock answer – 'education' – is
telling. All this does not mean that environmentalism is
dead, or that the Green revolution is over. What is usable in
the Green critique has largely been subsumed by the political system. What is not has been ejected. The activists will
continue to engage in adversarial confrontation. The children of the hippies, wanting to reclaim the land for the
people, will continue to tread the lanes of England. The
Green witches will continue their pray-ins. And even at
governmental level, where serious undertakings to control
pollution and change technology have been given, promises have become subject to the compromise and give and
take of the political process. There is as yet no agreement on
the carbon tax, originally pushed hard by the Netherlands,
while implementation of the Dutch National Environmental Plan has so far lagged behind the detailed targets and
objectives it sets. Enforcement remains the problem, even
in the consensual world of Dutch politics; how much more
a problem, then, in the non-consensual societies?

There is one profound difference between politics and
other disciplines. In politics you can create an idea out of
nothing; persuade people to believe in institutions and
principles. Belief confers reality: mutual belief confers legitimacy. Institutions feed back into society, and reinforce
myth. That is why politics should never be for the artist, the
thinker, or the would-be free mind. The environmentalist
myth is that benevolent intentions and statements of prin-

ciple will avoid the necessity for painful decisions. Documents like the Brundtland Report, statements of principle such as those issued after the Exxon-Valdez disaster, and the commitment to a sustainable economy, have attracted a remarkable degree of consensus among the great and the good. They have in particular attracted focused support from the EC, and from those European visionaries who see here both a crusade that will gain the EC popular support among the member states, and a means of enforcing hegemony over the nationally based legislation and sovereignty which has proved so plainly inadequate in enforcing pollution control. The momentum of environmentalism has left the national level and entered the international level. Even if environmental legislation is handed back to the EC member states, as Jacques Delors wanted to do to sweeten the Maastricht Treaty, the fact that the richest EC countries have the highest standards and most innovative legislation will ensure that they eventually enforce these standards on other countries, in order to protect their economic interests.

To avoid massive trade differentials, consensus at international level is necessary. It is a precondition of successful action. Dry economists like Lawrence Summers of the World Bank will argue that poor countries need to produce cheaply and market their unpolluted resources till they reach a ceiling equivalent to that of richer countries. In developing countries outside Europe this may happen for a while. However, in Central and Eastern Europe the importance of the environment in the Velvet Revolutions of 1989 will make it politically difficult to keep selling space for toxic waste disposal, or to buy pesticides no longer allowed in the West. The seething gossip that passes for political comment in these countries is alert to such events. If environmental ethics are taken seriously in the West, then they will be accepted in Eastern Europe as a standard that will have to be adopted. Not only is this necessary in order to join the EC (for those countries likely to do so within ten to fifteen years) but it is necessary to abide by product standards for exports. For some, environmental issues may be seen as yet another puzzling or irritating Western ploy, hypocritical, since in many Western countries the environment is a mess, but it will still be something to pay lip service to. And from acceptance of the ethic comes implementation.

Yet, in the process of rationalising environmentalism, of costing it, of playing trade wars with it, our concern for the intangible beauties of the natural world, our entirely 'speciesist' love of the most intelligent mammals, may go by the board.

We should salute the integrity and courage of the Green activists this century. For all the contradictions, dangers, possible evils and authoritarianism that could flow from their activities, they have stood out for the values that matter. But if one could return to earth a hundred years from now, would the result of their efforts be an idyllic and conserved nature? Or would it be a West further impoverished by the demands to share burdens with the developing world? If the latter, then the environmentalist will have failed. Because only the maligned Western world has the money and the will to conserve its environment. It is the 'Northern White Empire''s last burden, and may be its last crusade.

References

Abby, E. (1975) *The Monkey-Wrench Gang*, New York.

Ashworth, C.E. (1980) 'Flying Saucers, Spoon-Bending and Atlantis. A Structural Analysis of New Mythologies', *Sociological Review* 28, pp. 253–76.

Augstein, Rudolf (1989) *Historikerstreit*, Munich.

Baden, J. (1989) 'The New Environmentalists. Money, Politics, Nature', unpublished paper, March.

Bahro, R. (1979) *The Alternative in Eastern Europe*, New York.

Bahro, R. (1986) *Building the Green Movement*, London.

Balfour, E. (1941) *The Living Soil*, London.

Bergmann, K. (1970) *Agrarromantik und Großstadtfeindschaft*, Meisenheim am Glan.

Bloom, A. (1987) *The Closing of the American Mind*, New York.

Boulding, K. (1970) *Beyond Economics: Essays on Society, Religion and Ethics*, Ann Arbor, Michigan.

Bramwell, A. (1985) *Blood and Soil. R. Walther Darré and Hitler's 'Green Party'*, Bourne End

———, (20 November 1987) 'Widespread Seeds of the Green Revolution', *Times Higher Education Supplement*.

———, (1989) *Ecology in the Twentieth Century. A History*, London.

Buchanan, J.M. (1975) *The Limits of Liberty. Between Anarchy and Leviathan*, Chicago.

Buchanan, J.M. (1979) *What Should Economists Do?* Indianapolis.

Cairncross, F. (1991) *Costing the Earth*, London.

Callenbach, E. (1977) *Ecotopia*, New York. First published 1972.

Capra, F. (1988) *'Einfuehrung', Wendezeit*, Munich.

Carson, R. (1962) *Silent Spring*, New York.

Chisholm, A. (1972) *Conversations with Ecologists*, London.

Cotgrove, S. (1982) *Catastrophe or Cornucopia? The Environment, Politics and the Future*, Chichester.

Cotgrove, S. and Duff, A. (1981) 'Environmentalism, Values and Social Change', *British Journal of Sociology* 32, pp. 92–110.

Council on Environmental Quality (1982) *The Global 2000 Report to the President*, Washington D.C.

Dani, M. (1989) 'The Italian Ecological Movement: between Moderation and Radicalism', unpublished paper circulated at Modern History Seminar, Oxford University.

Davy, J. (1984) 'Dr E.F. Schumacher. An Appreciation', in S. Kumar (ed.), *The Schumacher Lectures*, Vol. II, pp. xi–xv.

Devall, B. (1990) *Simple in Means, Rich in Ends. Practising Deep Ecology*, London.

Dobson, A. (1990) *Green Political Thought*, London.

Driesch, H. (1914) *The History and Theory of Vitalism*, London.

Ebbatson, R. (1980) *Lawrence and the Nature Tradition*, Hassocks, Sussex.

Ehrlich, P.R. (1968) *The Population Bomb*, New York.

Ehrlich, P.R. and Ehrlich, A.H. (1970) *Population, Resources, Environment*, San Francisco.

Elkington, J. and Burke, T. (1987) *The Green Capitalists*, London.

Evans, R. (1989) *In Hitler's Shadow. West German Historians and the Attempt to Escape from the Nazi Past*, London.

Featherstone, Mike (June 1988) 'Post-Modernism', *Theory, Culture and Society*, Vol. 5, pp. 2–3.

Galtung, J. (1984) 'Sinking with Style', in S. Kumar (ed.), *The Schumacher Lectures*, London, Vol. II, pp. 1–22.

German Green Party Manifesto (1983) Berlin.

Gerschenkron, A. (1962) *Economic Backwardness in Economic Perspective*, Cambridge, Mass.

Goldsmith, E. (1972) *A Blueprint for Survival*, London.

Goodland, R.J.A. (1989) *Environment and Development: Progress of the World Bank (and Speculation towards Sustainability)* (World Bank Paper), Washington, DC.

Gore, Albert (1992) *Earth in the Balance: Forging a common purpose*, London.

Gould, P. (1988) *Early Green Politics*, Brighton, Sussex.

Gröning, G. and Wolschke-Bulmahn, J. (1987) 'Politics, Planning and the Protection of Nature. Political Abuse of Early Ecological Ideas in Germany, 1933–45', *Planning Perspectives* 2, pp. 127–48.

Grove, R. (1985) 'Incipient Conservationism in the Cape Colony and the Emergence of Colonial Environmental Policies in Southern Africa', unpublished paper presented at a seminar at Cambridge University.

Gruhl, H. (1975) *Ein Planet Wird Geplundert*, Frankfurt.

Habermas, J. (1985) *Der philosophische Diskurs der Moderne: Zwolf Vorlesungen*, Frankfurt.

Harvey, D. (1989) *The Condition of Post-Modernity. An Enquiry into the Conditions of Cultural Change*, Oxford.

Hele-King, D. (1989) 'All Kinds of Everything', *Times Literary Supplement*, 20–26 October.

Huelsberg, W. (1988) *The German Greens*, London.

Illich, I. (1971) *Deschooling Society*, London.

Inglehart, R. (1981) 'Post-Materialism in an Environment of Insecurity', *American Political Science Review* 85, pp. 880–900.

International Union for the Conservation of Nature (1980) *World Conservation Strategy*, Geneva.

Irvine, S. and Ponton, A. (1988) *A Green Manifesto. Policies for a Green Future*, London.

Jünger, F. (1975) *Bussauer Manifesto*, n.p.

Kahn, H. and Simon, J., eds (1984) *The Resourceful Earth. A Response to Global 2000*, Oxford.

Keith, W. (1975) *The Rural Tradition*, Hassocks, Sussex.

Kelly, P. (1984) *Fighting for Hope*, London.

Kruik, O. and Verbruggen, H. (1991) *In Search of Indicators of Economic Development*, Dordrecht.

Langhuth, G. (1986) *The Green Factor in German Politics. From Protest Movement to Political Party*, Boulder, Colorado.

Leopold, A. (1968) *The Sand Country Almanac*, Oxford. First published 1949.

Lowe, P. and Goyder, J. (1983) *Environmental Groups in Politics*, London.

Lowe, P. and Ruedig, W. (1986a) 'Review Article: Political Ecology and the Social Sciences – The State of the Art', *British Journal of Political Science* 16, pp. 513–50.

———, (1986b) 'The Withered "Greening" of British Politics. A Study of the Ecology Party', *Political Studies* 34, pp. 262–84.

Luhmann, N. (1989) *Ecological Communication*, London.

Lutzenberger, J. (1986) 'The Fate of the Rain-Forests', unpublished Schumacher Lecture, Bristol.

McBurney, S. (1990) *Ecology into Economics Won't Go. Or, Life is not a Concept*, Bideford, Devon.

McCormick, J. (1989) *The Global Environmental Movement*, London.

Maddox, J. (1972) *The Doomsday Syndrome*, London.

Marcuse, H. (1964) *One Dimensional Man: studies in the ideology of advanced industrial society*, London.

———, (1966) *Eros and Civilization: a philosophical enquiry into Freud*, Boston.

Marsh, J. (1982) *Back to the Land. The Pastoral Impulse in Victorian England, 1880–1914*, London.

Martinez-Alier, J. with Schluppmann, Klaus (1987) *Ecological Economics*, Oxford.

Meadows, D.H. *et al.* (1972) *The Limits to Growth*, New York.

Medawar, P. (1982) *Pluto's Republic*, Oxford.

Merchant, C. (1980) *The Death of Nature. Women, Ecology and the Scientific Revolution*, New York.

Mishan, E.J. (1967) *The Costs of Economic Growth*, London.

———, (1969) *Growth: the price we pay*, London.

Mohler, A. (1978) *Der Traum vom Naturparadies. Anmerkungen zur Oekologischen Gedankengut*, Munich.

Naess, A. (1989) *Ecology, Communication and Lifestyle*, Cambridge.

Nash, R. (1989) *The Rights of Nature. A History of Environmental Ethics*, Madison, Wisconsin.

Nelson, R. (1991) *Modern Economic Theology* (publication forthcoming).

OECD Environment Secretariat (1989) *The Concept of Sustainable Development and its Practical Economic Implications*, Paris.

Oelschlaeger, M. (1991) *The Idea of Wilderness*, London and New Haven, Conn.

Paehlke, R. (1989) *Environmentalism and the Future of Progressive Politics*, New Haven, Conn.

Papadakis, E. (1984) *The Green Movement in Germany*, London.

Parkin, S. (1989) *Green Politics. An International Guide*, London.

Pearce, D. *et al.* (1989) *Blueprint for a Green Economy*, London.

Pepper, D. (1985) *The Roots of Modern Environmentalism*, London.

Phillips, D. (ed.) (1983) *German Universities after the Surrender*, Oxford.

Pois, R. (1985) *National Socialism and the Religion of Nature*, London.

Popper, K. (1982) *Unended Quest. An Intellectual Biography*, London.

Porritt, J. (1986) *Seeing Green*, Oxford.

Postrel, V. (1990) 'The Green Road to Serfdom', *Reason*, April.

Regan, T. (1988) *Mammalian Rights*, London.

Reich, C. (1971) *The Greening of America*, London.

Repetto, R. (1989) *Wasting Assets. The Need for Natural Resource Accounting* (World Resources Institute Paper), Washington, DC.

Rifkin, J. with Howard, T. (1980) *Entropy. A New World View*, London.

RIVM (Dutch National Institute of Public Health and Environmental Protection) (1991) *Concern for Tomorrow 1985–2010*, Amsterdam.

Roderick, R. (1986) *Habermas and the Foundations of Critical Theories*, London.

Rostow, W. (1960) *The Stages of Economic Growth. A Non-Communist Manifesto*, Cambridge.

Sale, K. (1984) 'Mother of All. An Introduction to Bio-regionalism', in S. Kumar (ed.), *The Schumacher Lectures*, Vol. II, London.

Sale, K. (1985) *Dwellers in the Land. The Bioregional Vision*, San Francisco.

Schumacher, E.F. (1973) *Small is Beautiful*, London.

Seabrook, J. (1990) *The Myth of the Market. Promises and Illusions*, Bideford, Devon.

Singer, P. (1976) *Animal Liberation*, London.

Singer, S. Fred (1984) 'World Demand for Oil', in H. Kahn and J. Simon (eds), *The Resourceful Earth*, Oxford.

Smith, M. (1988) *Pacific Visions. Californian Scientists and the Environment, 1815–1900*, New Haven and London.

Spretnak, C. and Capra, F. (1985) *Green Politics*, London.

Stapledon, G. (1964) *Human Ecology*, ed. R. Waller, London.

Thayer, G. (1965) *The British Political Fringe*, London.

Trommer, G. (1990) *Natur im Kopf. Die Geschichte oekologisch bedeutsamer Naturvorstellungen in deutschen Bildungskonzepten*, Weinheim.

Vereker, C. (1967) *Eighteenth-Century Optimism*, Liverpool.

Walker, M. (17 August 1990) 'Shock Troops in the Eco-War', *The Guardian*.

Ward, B. and Dubos, R. (1972) *Only One Earth. The Care and Maintenance of a Small Planet*, London.

Weindling, P. (1989) *Health and German Politics between National Unification and Nazism 1870–1975*, Cambridge.

Wohl, R. (1980) *The Generation of 1914*, London.

Wolpert, L. (1993) *The Unnatural Nature of Science*, London.

World Commission on Environment and Development (Brundtland Commission) (1987) *Our Common Future*, Oxford and New York.

Worster, D. (1977) *Nature's Economy. The Roots of Ecology*, Cambridge.

Index